Geometrical Kaleidoscope

Second Edition

Problem Solving in Mathematics and Beyond

Print ISSN: 2591-7234
Online ISSN: 2591-7242

Series Editor: Dr. Alfred S. Posamentier
Distinguished Lecturer
New York City College of Technology - City University of New York

There are countless applications that would be considered problem solving in mathematics and beyond. One could even argue that most of mathematics in one way or another involves solving problems. However, this series is intended to be of interest to the general audience with the sole purpose of demonstrating the power and beauty of mathematics through clever problem-solving experiences.

Each of the books will be aimed at the general audience, which implies that the writing level will be such that it will not engulfed in technical language — rather the language will be simple everyday language so that the focus can remain on the content and not be distracted by unnecessarily sophiscated language. Again, the primary purpose of this series is to approach the topic of mathematics problem-solving in a most appealing and attractive way in order to win more of the general public to appreciate his most important subject rather than to fear it. At the same time we expect that professionals in the scientific community will also find these books attractive, as they will provide many entertaining surprises for the unsuspecting reader.

Published

For the complete list of volumes in this series, please visit www.worldscientific.com/series/psmb

Problem Solving in
Mathematics and Beyond

Volume 33

Geometrical Kaleidoscope

Second Edition

Boris Pritsker

We World Scientific

NEW JERSEY · LONDON · SINGAPORE · BEIJING · SHANGHAI · HONG KONG · TAIPEI · CHENNAI · TOKYO

Published by

World Scientific Publishing Co. Pte. Ltd.
5 Toh Tuck Link, Singapore 596224
USA office: 27 Warren Street, Suite 401-402, Hackensack, NJ 07601
UK office: 57 Shelton Street, Covent Garden, London WC2H 9HE

Library of Congress Cataloging-in-Publication Data
Names: Pritsker, Boris, author.
Title: Geometrical kaleidoscope / Boris Pritsker.
Description: Second edition. | New Jersey : World Scientific, [2024] |
    Series: Problem solving in mathematics and beyond, 2591-7234 ; 33 |
    Includes bibliographical references and index.
Identifiers: LCCN 2023049057 | ISBN 9789811285271 (hardcover) |
    ISBN 9789811285608 (paperback) | ISBN 9789811285288 (ebook for institutions) |
    ISBN 9789811285295 (ebook for individuals)
Subjects: LCSH: Geometry. | BISAC: MATHEMATICS / Geometry.
Classification: LCC QA445 .P75 2024 | DDC 516--dc23/eng/20231117
LC record available at https://lccn.loc.gov/2023049057

British Library Cataloguing-in-Publication Data
A catalogue record for this book is available from the British Library.

For any available supplementary material, please visit
https://www.worldscientific.com/worldscibooks/10.1142/13650#t=suppl

Desk Editor: Nimal Koliyat/Tan Rok Ting

Typeset by Stallion Press
Email: enquiries@stallionpress.com

Preface

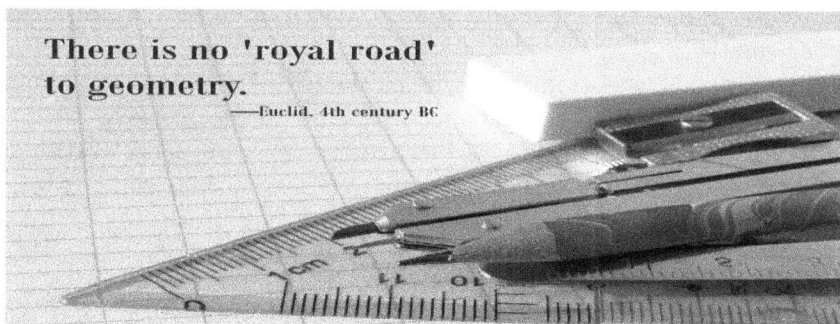

There is no 'royal road' to geometry.
——Euclid, 4th century BC

"I can't do it anymore!" My oldest son Alex put aside his geometry homework. "And please don't tell me I would ever use this nightmare somewhere, somehow in my life!" My heroic attempts to explain to him the solution of the problem involving the properties of the bisector of a triangle failed miserably. "You should be ashamed of yourself. I recently published an article about the properties of the bisector (see Chapter 4) and you don't have any patience even to listen to me." I made this final argument trying to get his attention to the subject. All of a sudden there was a spark in his eyes and he looked at me with a definite expression of interest on his face.

"May I see it?"

"Sure, here it is." I was very proud of myself.

He looked at the article, read a few paragraphs, and then patiently let me finish the explanation of the solution to his homework problem.

"Could you please make a copy of this article for me?" he asked when we finished working on the homework. It was a real pleasure to hear something like this from him. Obviously, the kid wanted to show the copy to his school friends and gain some popularity because of his father's math skills.

In a few days, he got back from school and told me about getting an A on his most recent geometry test. My efforts had finally paid off and I was very happy for a few days. Then I met his school friend. He told me the true story about how an A was earned.

Alex is a very good tennis player. His interests have always been in sports and he has never been a great math student. He has compensated for his lack of knowledge and desire to learn the subject by making jokes and telling funny stories during the classes. Teachers always had a tough time trying to keep him quiet. By the way, he never mentioned in a classroom that his dad used to be a professional mathematician.

That day he was more annoying to the teacher than usual. At the end of the session, Alex announced that he did not have to do the assigned homework, because he just studied those properties with his dad while reading his published article in a math magazine. It was obviously hard to imagine that he was telling the truth.

"I am sick and tired of your fantasies. I'll teach you a lesson this time. If you bring that article in the class tomorrow, you'll get an A on the next test, without even taking it. But if you don't have it, your homework assignment will be tripled and you would need to hand in every single problem to me," the teacher said.

The teacher's desire to make fun of this annoying kid was quite understandable.

You know the end of the story. The teacher kept his word.

That was my son's only experience of using geometry in real life so far. I have to admit, that even though he did not succeed in geometry class, he did apply the methods I was teaching him. How about the auxiliary element introduction (see Chapter 8) in achieving the goal of getting a good grade without making a real effort? He used my publication as an auxiliary element to provoke the teacher. The problem of getting a good grade was solved without going deeply into the material, even without taking the test.

Many years passed. Alex is a grown-up man and has his own family now. When both of us recall this funny story, he regrets not devoting more time to math topics and admits the undeniable benefits of studying geometry in developing and strengthening logical reasoning.

There is no need to mention the importance of mathematics, including geometry, in early childhood education. However, math is often the most difficult subject for many children, and they sometimes lose interest and motivation.

I hope that the methods and concepts outlined and discussed in the book will stimulate readers to explore unfamiliar or little-known aspects of geometry. My aim was to extend an invitation to geometry to those with a passion for and admiration of the great world of mathematics, including my youngest son Bryan, who loves math and just graduated magna cum laude (very proud of him!) from college with BA in mathematics. His interest in mathematics was one of the factors that inspired this book. He was also a big help in editing a few chapters, and I am grateful to him for doing that.

Usually, we experience the greatest difficulties at the beginning of a problem-solving process. Where should you start? How in the world can you link together some of the conditions of the problem? I think the answer to these typical questions is to "recognize" the problem immediately after seeing it. By "recognize" I mean the ability to identify its type and determine the most important of the given conditions to use as bricks in building a logical chain. This book focuses on geometric thinking — what it means, how to develop it, and how to recognize it.

Plato, one of the greatest philosophers in human history, displayed the slogan "Let no one ignorant of geometry enter here" in front of his Academy.

The goal of this book is to provide insight into some enjoyable and fascinating aspects of geometry and to reveal many interesting geometrical properties. The topics cover well-known properties and theorems, including classic examples such as Archimedes' Law of the Lever, the Pythagorean Theorem, Heron's formula, Brahmagupta's formula, Appollonius's Theorem, Euler's line properties, the Nine-Point Circle, Fagnano's Problem, the Steiner-Lehmus Theorem, Napoleon's Theorem, Ceva's Theorem, Pompeiu's Theorem, and Morley's Miracle. "Geometrical Kaleidoscope" consists of a kaleidoscope of topics that seem to not be related at first glance. However, that perception disappears as you go from chapter to chapter and explore the multitude of surprising relationships, unexpected connections, and links. The greatest attention should be given to the underlying principles and the major steps in solving problems.

Readers solving a chain of problems will learn from them general techniques, rather than isolated instances of the application of a technique. They might use the problems as a nucleus around which to build a set of more difficult problems, which yield the same method of solution. Also, we demonstrate how the search for multiple solutions to a problem helps in getting important generalizations and new unexpected results.

I believe that the best way to learn mathematics is to *do* mathematics.

Marcus du Sautoy, well-known for his work popularizing mathematics, said, "The power of mathematics is often to change one thing into another, to change geometry into language." I hope this book will offer another small step in that direction and that the topics and problems present an enjoyable experience for readers.

I am very grateful to the readers who reviewed the book on Amazon's website and LinkedIn and shared with me their views. Their constructive criticism and encouraging comments were important in my decision to update and revise the 2017 edition of the book and prepare this new edition for publication.

This book is dedicated to both of my sons, Bryan and Alex with his family, with special dedication going to my adorable granddaughters Liana and Luna. I love them very much and wish to see them all happy and successful in achieving their goals in life. It is also dedicated to my wife Irina, who almost did not complain and patiently survived the lack of my attention while I was working on the book. It is dedicated finally to my late parents, whose love, support, and inspiration always helped me in the past, and will stay with me for the rest of my life.

About the Author

Boris Pritsker was born in Kiev, Ukraine. He studied mathematics at Kiev State Pedagogical University, where he graduated summa cum laude with the US equivalent of a B.S. degree in math/education. Pritsker worked as a math teacher in high schools including a special math-oriented school for gifted and talented students with advanced programs in algebra, geometry, trigonometry, and calculus. He developed educational seminars for studying classic math problems encouraging students to look for alternative solutions. He also trained school math teams for participation in math competitions; several students from these math teams won the prestigious local and regional math Olympiad contests.

After immigrating to the United States, Pritsker earned an MBA degree magna cum laude from the Graduate School of Baruch College, City University of New York. Pritsker is a licensed CPA in New York State. For the past 25 years, he has been employed by CBIZ Marks Paneth LLC, New York City accounting and consulting firm, where he became a director.

Never leaving behind his curiosity and devotion to mathematical education development and research, through the years Pritsker has been exploring various challenging and interesting topics in Euclidean geometry, algebra, trigonometry, and pre-calculus which he offered for publication in math educational journals. He has published problems and articles in the Soviet Union, USA,

Singapore, and Australian magazines such as "Математика в школе" (in Russian), (*Mathematics in School*), *NY State Mathematics Teachers' Journal, Quantum, New England Mathematics Journal, Mathematics and Informatics, Journal of Recreational Mathematics, Mathematics Teacher, Mathematics Competitions* (journal of the World Federation of National Mathematics Competitions).

Pritsker authored four internationally acclaimed mathematics books *Geometrical Kaleidoscope*, Dover Publications, 2017 (2nd edition of which is this title), *The Equations World*, Dover Publications, 2019, *Mathematical Labyrinths. Pathfinding*, World Scientific, 2021, and *Expanding Mathematical Toolbox: Interweaving Topics, Problems, and Solutions*, Chapman & Hall/CRC, 2023.

Contents

Chapter 1

Medians, Centroid, and Center of Gravity of a System of Points

In every triangle, there are three important special segments: medians, altitudes, and angle bisectors. We will examine some interesting properties of them and will demonstrate their applications in problem-solving. We start with the medians of a triangle.

A median of a triangle is a segment drawn from a vertex to the midpoint of the opposite side.

The three medians in a triangle are concurrent and their common point is called the centroid of the triangle. It is $\frac{2}{3}$ of the way from a vertex to the opposite midpoint. That is, the medians of a triangle divide each other in the ratio 2:1.

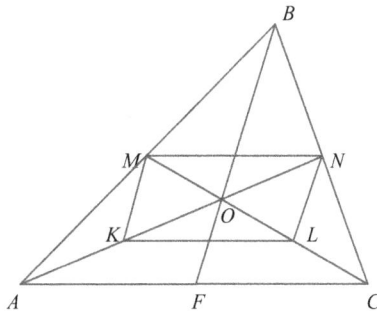

Let us prove the above statements.

In triangle ABC, AN, BF, and CM are the medians. We need to prove that they pass through point O and O splits each median in the ratio 2:1.

Triangles MBN and ABC are similar because they have a common angle B and the respective sides that form angle B are in the same ratio: $MB = \frac{1}{2}AB$, and $BN = \frac{1}{2}BC$.

Therefore alternate interior angles are congruent, $\angle BMN = \angle BAC$ and $\angle BNM = \angle BCA$ as respective angles of the similar triangles, and accordingly, $MN \parallel AC$. If we select points K and L on OA and OC respectively such that $OK = KA$ and $OL = LC$, then by similar reasoning $KL \parallel AC$ (triangles KOL and AOC are similar). It follows that $MN = KL = \frac{1}{2}AC$ and $MN \parallel KL$ (both MN and KL are parallel to AC). Thus $KMNL$ is a parallelogram. O is the point of intersection of its diagonals, which divides the diagonals in half. Then $ON = OK$ and $OM = OL$. Because we've chosen K and L such that $OK = KA$ and $OL = LC$, we obtain that $AK = KO = ON$ and $CL = LO = OM$, so point O trisects the medians AN and CM. Since we could equally well have begun with another pair of medians, the three medians are concurrent and the point of their intersection divides each median in the ratio 2:1. The proof is completed.

A powerful property evolved during the proof:

The midsegment or midline of a triangle (the segment that connects two midpoints of two sides) is parallel to the opposite side of a triangle and is half as long.

The triangle formed by joining the midpoints of the sides of a triangle is called the *medial* triangle. As we observed, the medial triangle has three sides parallel to the sides of the original triangle, so the triangles are similar. The ratio of the respective sides equals $\frac{1}{2}$.

Problem 1. There are three non-collinear points A, B, and K. Construct the triangle ABC, such that point K is its centroid and points A and B are vertices.

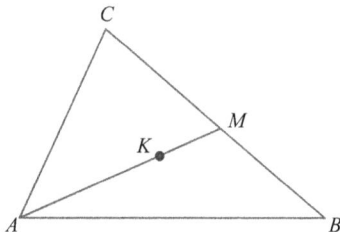

Solution. For the solution, it suffices to notice that if K is the centroid of the triangle ABC, then by locating the point M on the extension of AK such that $MK = \frac{1}{2}AK$, we get the midpoint of side BC. By drawing BM and locating point C on its extension such that $MC = MB$, we get the third vertex of the triangle. The last step is to connect points A and C. ABC is the desired triangle.

Our constructions are performed with only a compass and straightedge (a ruler without markings to draw straight lines), unless it is specifically indicated otherwise. The explanations of basic constructions will be omitted (as we omitted them in the problem above). However, we recommend that readers thoroughly perform every step in a construction process and its justification, determine how unique the solution is, and review the alternative solutions.

The medians of a triangle have a few important properties.

Theorem 1 (Appollonius's Theorem). *The sum of squares of any two sides of any triangle equals twice the sum of a square of half the third side and the square of the median bisecting the third side.*

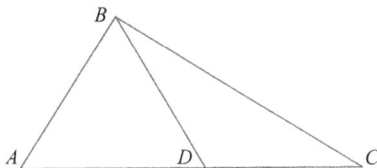

Proof.   If BD is the median in triangle ABC, then we have to prove that

$$AB^2 + BC^2 = 2(BD^2 + AD^2).$$

Let's denote $\angle ADB = \alpha$ and $\angle BDC = \beta$. Angles α and β are supplementary. Thus $\beta = 180° - \alpha$ and so

$$\cos \beta = \cos(180° - \alpha) = -\cos \alpha \tag{1}$$

Now we will apply the Law of cosines to the triangles ABD and BDC:

$$AB^2 = BD^2 + AD^2 - 2AD \cdot BD \cdot \cos \alpha,$$
$$BC^2 = BD^2 + CD^2 - 2CD \cdot BD \cdot \cos \beta.$$

Add those equalities. Recalling that $AD = DC$ (since BD is a median) and substituting (1) into the second equality, we obtain that $AB^2 + BC^2 = 2(BD^2 + AD^2)$, as required.

If we denote the sides of the triangle ABC as $AB = c$, $BC = a$, $AC = b$ and let the median $BD = m_b$, then after simple manipulations the expression of a median of a triangle in terms of its sides is

$$m_b = \frac{1}{2}\sqrt{2a^2 + 2c^2 - b^2}.$$

Similar expressions hold for the other medians:

$$m_a = \frac{1}{2}\sqrt{2b^2 + 2c^2 - a^2},$$

$$m_c = \frac{1}{2}\sqrt{2a^2 + 2b^2 - c^2}.$$

It should not be difficult to use the above formulas to express each side of a triangle in terms of its medians. We invite readers to perform those manipulations and memorize the formulas as important and efficient tools for future use:

$$b = \frac{2}{3}\sqrt{2m_a^2 + 2m_c^2 - m_b^2},$$

$$a = \frac{2}{3}\sqrt{2m_b^2 + 2m_c^2 - m_a^2},$$

$$c = \frac{2}{3}\sqrt{2m_a^2 + 2m_b^2 - m_c^2}.$$

Theorem 2. *Each median of a triangle divides the area of a triangle in half.*

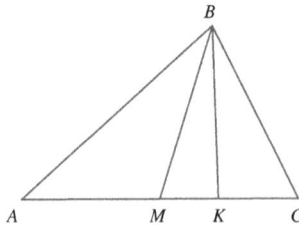

Proof. BM is the median and BK is the altitude in the triangle ABC. We have to prove that the areas of the triangles ABM and BMC are equal.

For the proof, we will apply the formula for the area of a triangle $S = \frac{1}{2}ah$ (a is the base, and h is an altitude dropped to that base) and notice that triangles AMB and MBC have common altitude BK and equal bases $AM = MC$.

A direct corollary is:

Theorem 3. *A triangle is partitioned by its medians into six triangles of equal area.*

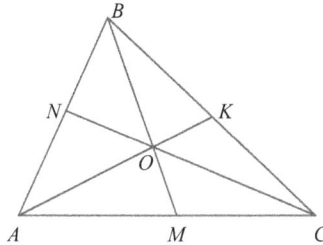

Proof. The proof becomes obvious if we compare the equal areas of triangles ABM and CBM as areas of the triangles in which they are partitioned by the medians.

From the equalities $S_{AMB} = S_{AOB} + S_{AOM}$ and $S_{BMC} = S_{BOC} + S_{COM}$ it follows that

$$S_{AOB} + S_{AOM} = S_{BOC} + S_{COM}. \tag{1}$$

Note that OM, ON, and OK (O is the centroid) are the medians in the triangles AOC, AOB, and BOC, respectively. Then $S_{AOM} = S_{COM}$, $S_{AON} = S_{BON}$, and $S_{BOK} = S_{COK}$.

We see that (1) yields $S_{AOB} = S_{BOC}$, or $2S_{BON} = 2S_{BOK}$ and finally, $S_{BON} = S_{BOK}$.

Similarly, it is easy to show that $S_{KOC} = S_{MOC}$. We see that the areas of each of the six small triangles are equal, and each area is $\frac{1}{6}$ of the area of the big triangle ABC.

The centroid of a triangle has another commonly used name — center of mass, or center of gravity. Before explaining the reason for that second name, we will introduce the general definition of center of mass for any system of material points.

M. Balk and V. Boltyanski wrote an excellent article "Utilization of the center of gravity at the math facultative studies" in *Matematika v shcole*, 2, 1984, pp. 45–50, (**М.Б. Бланк, В.Г. Болтянский,**

"Применение понятия центра   масс на   факультативных и кружковых занятиях", *Математика в школе*, 2, 1984, с. 45–50, in Russian) in which they covered the major applications of various mechanical laws in geometry. We restrict our attention here to only a few techniques closely related to the application of the properties of the center of gravity of the system of points in problem-solving.

In physics, a material point is defined as an object the size of which is negligible in comparison with the distances in the problem. For simplicity, such an object is considered just as a point (it is assumed that its whole weight is concentrated at this one point). If the mass m is concentrated at point A, then we will denote this material point as mA. By definition, point Z would be the center of mass of the system of material points $m_1A_1, m_2A_2, \ldots, m_nA_n$, if sum of all the respective vectors equals the 0 vector: $m_1\overrightarrow{ZA_1} + m_2\overrightarrow{ZA_2} + \cdots + m_n\overrightarrow{ZA_n} = \vec{0}$. The center of gravity is the point at which the whole system would be perfectly balanced, assuming uniform density and a uniform gravitational field.

To better understand the following material, let's agree to translate the words "concentrate a mass m at the point A" as "relate the number m to the point A". The expression "material point mA" should then mean "point A along with a number m, associated with point A". The number m we will call "the mass of a material point mA". Let's now prove the basic theorem about the center of gravity of a system of material points.

Theorem 4. *If a point Z is the center of gravity (center of mass) of a system of material points $m_1A_1, m_2A_2, \ldots, m_nA_n$, then for any point O the following vector equality will hold true:*

$$\overrightarrow{OZ} = \frac{m_1\overrightarrow{OA_1} + m_2\overrightarrow{OA_2} + \cdots + m_n\overrightarrow{OA_n}}{m_1 + m_2 + \cdots + m_n}. \qquad (1)$$

Proof. To prove the statement of the theorem, we will use the following property of the vectors: for any points O, A, and Z, vector ZA might be expressed as the difference of vectors $\overrightarrow{OA}$ and $\overrightarrow{OZ}$, $\overrightarrow{ZA} = \overrightarrow{OA} - \overrightarrow{OZ}$. When each vector is multiplied by m, that becomes:

$$m\overrightarrow{ZA} = m(\overrightarrow{OA} - \overrightarrow{OZ}). \tag{2}$$

Now we turn to the definition of the center of gravity of a system of material points $m_1A_1, m_2A_2, \ldots, m_nA_n$. If Z is the center of gravity of a system of material points $m_1A_1, m_2A_2, \ldots, m_nA_n$, then

$$m_1\overrightarrow{ZA_1} + m_2\overrightarrow{ZA_2} + \cdots + m_n\overrightarrow{ZA_n} = \vec{0}.$$

By substituting (2) for each vector, we get that

$$m_1(\overrightarrow{OA_1} - \overrightarrow{OZ}) + m_2(\overrightarrow{OA_2} - \overrightarrow{OZ}) + \cdots + m_n(\overrightarrow{OA_n} - \overrightarrow{OZ}) = \vec{0}.$$

A few steps of fairly simple manipulations yield the desired result:

$$\overrightarrow{OZ} = \frac{m_1\overrightarrow{OA_1} + m_2\overrightarrow{OA_2} + \cdots + m_n\overrightarrow{OA_n}}{m_1 + m_2 + \cdots + m_n}.$$

The reverse statement is also correct: if for any point O the equality (1) holds, then Z is the center of gravity of the system of material points $m_1A_1, m_2A_2, \ldots, m_nA_n$.

We invite readers to prove this assertion on their own.

According to the theorem, any system of material points must have a center of gravity, which is the one and only point defined by (1). The center of mass is the arithmetic mean of the points weighted by their weights.

The formula (1) is much simplified when the total mass of the system is 1. Then we have a distribution of the mass 1 among points $A_1, A_2, \ldots, A_n$. If $x_1A_1, x_2A_2, \ldots, x_nA_n$ are the material points with mass 1 distributed evenly among them and Z is the center of gravity of the system, then for any point O we get the simplified expression:

$$\overrightarrow{OZ} = x_1\overrightarrow{OA_1} + x_2\overrightarrow{OA_2} + \cdots + x_n\overrightarrow{OA_n}.$$

What would happen with the center of gravity of a system of material points if the points are regrouped or reorganized in some way? That is an important practical matter to understand and apply in the solution of many problems. The following theorem gives an answer.

Theorem 5. *Assume there is a system of k material points $m_1A_1, m_2A_2, \ldots, m_kA_k$ in the total system of n material points $m_1A_1, m_2A_2, \ldots, m_nA_n$. Let C be the center of gravity of the system of k points. If the weight of the system of k points is to be concentrated at point C, then the location of the center of gravity of the whole system of n points will not change. In other words, the system of material points $(m_1 + m_2 + \cdots + m_k)C, m_{k+1}A_{k+1}, \ldots, m_nA_n$ and the original system of n points $m_1A_1, m_2A_2, \ldots, m_nA_n$ have the same center of gravity.*

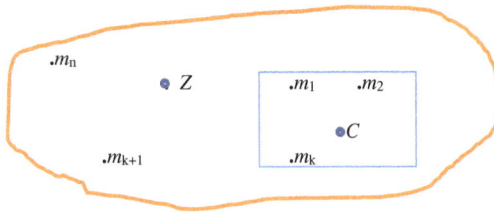

Proof. We restrict our consideration for the case $n = 5$ and $k = 3$ (a system of 3 points is apportioned in the total system of 5 points). The general proof would be exactly the same.

Let Z be the center of gravity of system of 5 points $m_1A_1, m_2A_2, m_3A_3, m_4A_4, m_5A_5$.

Then

$$m_1\overrightarrow{ZA_1} + m_2\overrightarrow{ZA_2} + m_3\overrightarrow{ZA_3} + m_4\overrightarrow{ZA_4} + m_5\overrightarrow{ZA_5} = \vec{0}. \qquad (3)$$

It was given that C is the center of gravity of three points m_1ZA_1, m_2ZA_2, and m_3ZA_3.

Therefore, $\overrightarrow{ZC} = \frac{m_1\overrightarrow{ZA_1} + m_2\overrightarrow{ZA_2} + m_3\overrightarrow{ZA_3}}{m_1 + m_2 + m_3}$, from which $m_1\overrightarrow{ZA_1} + m_2\overrightarrow{ZA_2} + m_3\overrightarrow{ZA_3} = (m_1 + m_2 + m_3)\overrightarrow{ZC}$. After substituting this expression into (3), we obtain that $(m_1 + m_2 + m_3)\overrightarrow{ZC} + m_4\overrightarrow{ZA_4} + m_5\overrightarrow{ZA_5} = \vec{0}$, which implies that by definition, point Z remains the center of gravity of the points $(m_1 + m_2 + m_3)C, m_4A_4$, and m_5A_5. This completes the proof of the theorem.

When we have the system of two material points m_1A_1 and m_2A_2, then by the definition, point Z is the center of gravity of such a system if $m_1\overrightarrow{ZA_1} + m_2\overrightarrow{ZA_2} = \vec{0}$, or $m_1\overrightarrow{ZA_1} = -m_2\overrightarrow{ZA_2}$.

The last vector equality implies that vectors ZA_1 and ZA_2 are collinear and have opposite directions, which means that Z lies on the segment A_1A_2 and the distance m_1ZA_1 equals the distance m_2ZA_2. This fact proves **Archimedes' Law of the Lever**. We see that $ZA_1/ZA_2 = m_2/m_1$, which yields that the center of gravity of two material points m_1A_1 and m_2A_2 divides the segment A_1A_2 in the ratio m_2/m_1 counting from the point A_1.

The prominent inventor and mathematician from ancient Greece, Archimedes (287–212 BC) was the first to introduce the method of discovering new geometrical theorems by the use of mechanics. He once said, "Give me place to stand and rest my lever on, and I can move the Earth." What place did he mean? What is the best place to stand to be in such position? Should it be the point of a perfect balance? If we talk about a triangle, then such point is its centroid. Now it is clear why it bears another name — *center of mass* or *center of gravity*.

In his work *On the Equilibrium of Planes* the great scholar formulated the statement, which we described above and which is well-known today as **Archimedes' Law of the Lever**:

Magnitudes are in equilibrium at distances reciprocally proportional to their weights.

The Law of the Lever

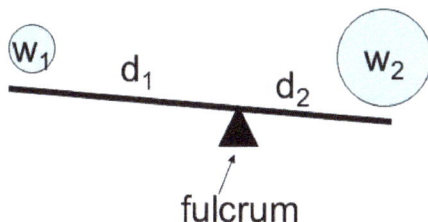

$$\mathbf{w_1 \times d_1 = w_2 \times d_2}$$

At this point, we are ready to entertain another vivid and elegant proof (its discovery is attributed to Archimedes) of the centroid's existence and its location in a triangle, which is at two-thirds the distance from a vertex to the midpoint of the opposite side.

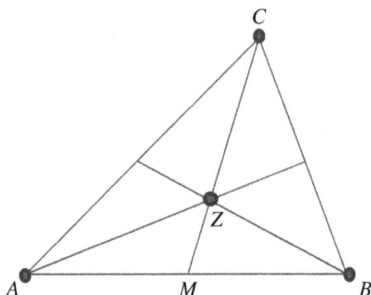

Let's put the same weight of 1 pound at each vertex of a triangle ABC. Denote by Z the center of gravity of this system of three material points $1A$, $1B$, and $1C$. If we shift the weights from points A and B to the midpoint M of the segment AB, then 2 pounds will be located at point M and still 1 pound at point C. That shift would not affect the location of the center of gravity Z. However, now our system consists of two material points $2M$ (with 2 pounds) and $1C$ (1 pound). Therefore, point Z must lie on CM. With no loss of generality, we could have selected another median and made exactly the same observation, hence Z must lie on each of the other two medians as well. According to Archimedes' Law of the Lever, if Z is the center of gravity of points $2M$ and $1C$, the distance from Z to M is half of the distance from Z to C, which proves the desired statement.

Applying the above principles, one can get unexpected elegant and beautiful solutions to many non-trivial problems. Here are a few examples.

Problem 2. For any convex hexagon, prove that the centroids of two triangles formed by joining the midpoints of the hexagon's non-adjacent sides coincide.

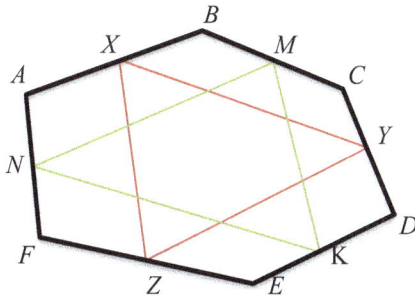

Solution. In the given hexagon $ABCDEF$, points X, M, Y, K, Z, and N are the midpoints of its sides. By connecting the points X, Y, and Z, we get the triangle XYZ; by connecting the points N, M, and K we get the triangle NMK. Our goal is to prove that the points of intersections of the medians of both triangles coincide. The problem might become a real disaster if you are not familiar with the definition of the center of gravity of a system of points and its properties. On the other hand, if you apply your knowledge to this situation, you get a simple and short solution without even locating the centroid of each triangle in the diagram. To solve this problem, we just need to find the location of the center of mass of the polygon by two different methods and compare our observations.

Let's put the same unit mass at each vertex of the hexagon. Then the center of mass of points A and B is concentrated at midpoint X of segment AB. The same is true for the midpoints of the other 5 sides of the hexagon. Obviously, the center of mass of the six vertices of the hexagon is the same point as the center of mass of the system of three points X, Y, and Z, **or** points N, K, and M. In other words, the center of mass of the hexagon has to be the same as the center of mass of triangle XYZ **or** triangle NKM. For each triangle, the center of mass is its centroid. Since the polygon has only one center of gravity, the centroids of the triangles XYZ and NKM must coincide. The proof is completed.

Problem 3. Given a square $ABCD$ with the side of a length a. What is the distance from a point P to the center of this square, if P satisfies $\overrightarrow{PA} + 3\overrightarrow{PB} + 3\overrightarrow{PC} + \overrightarrow{PD} = \vec{0}$?

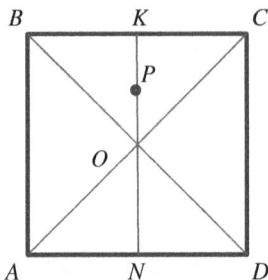

Solution. By the definition of the center of mass of a system of material points, point P has to be the center of mass of the material points $1A$, $3B$, $3C$, and $1D$. Also, note that O, the center of the square, is its center of mass. Let's shift the weight of 3 units from points B and C to their center of mass, the midpoint K of segment BC. Then the weight of 6 units is now concentrated at the point K, or we get the material point $6K$. Similarly, we shift the 1 unit weight from A and D to their center of mass, the midpoint N of the segment AD. We get the material point $2N$. Obviously, all three points K, N, and O must lie on the same straight line. As it was proved in Theorem 5, the location of the center of gravity of the system $1A$, $3B$, $3C$, and $1D$ will not change. However, now we will look at P as a center of gravity of two material points $6K$ and $2N$, which implies that P must lie on KN as well. By Archimedes' Law of the Lever, $\frac{PK}{PN} = \frac{2}{6}$. Then $PN = 3PK$.

Notice that $PK + PN = KN = a$ (as long as the length of the side of the square is a, then $KN = a$). We get that $PK = \frac{1}{4}a$. The final step is to observe that

$$PO = KO - PK = \frac{1}{2}a - \frac{1}{4}a = \frac{1}{4}a.$$

As you can see, the methods suggest the extensive use of vector algebra. The solutions of many problems involving the center of gravity may be significantly simplified with the introduction of vectors and manipulations with them. Very often, a problem solved by the application of the properties of the center of gravity may be translated into vector algebra and vice versa.

Here are a few exercises for your own practice.

Problem 4. BD and CE are the medians in a triangle ABC. M is its centroid. Prove that the area of the triangle BCM equals to the area of the quadrilateral $AEMD$.

Problem 5. Let m_a, m_b, and m_c be the medians of a triangle ABC. Denote by q the semi-sum of the medians: $q = \frac{1}{2}(m_a + m_b + m_c)$. Prove that the area S of the triangle ABC can be calculated by the formula $S = \frac{4}{3}\sqrt{q(q - m_a)(q - m_b)(q - m_c)}$.

Problem 6. In the given triangle ABC, m_1, m_2, and m_3 represent the lengths of its three medians, and $m_1^2 + m_2{}^2 = 5m_3^2$. Prove that ABC is a right triangle.

Problem 7. Point M lies on the side AC of the triangle ABC and $MC = 2AM$. Point N lies on the extension of side BC and $BN = BC$. Point P is the point of intersection of NM and AB. What is the ratio $AP{:}PB$?

Problem 8. The lengths of the sides of a triangle are 11, 13, and 12. Find the length of the median dropped to the longest side.

Chapter 2

Altitudes and the Orthocenter of a Triangle and Some of Its Properties

An altitude of a triangle is a line segment dropped through a vertex that is perpendicular to the opposite side. In this chapter, we will investigate some interesting properties of altitudes and the point of their intersection, called the *orthocenter* of a triangle. In Figure 1, AK, BM, and CN are the altitudes ($AK \perp BC$, $BM \perp AC$, and $CN \perp AB$) and H is the orthocenter of the triangle ABC.

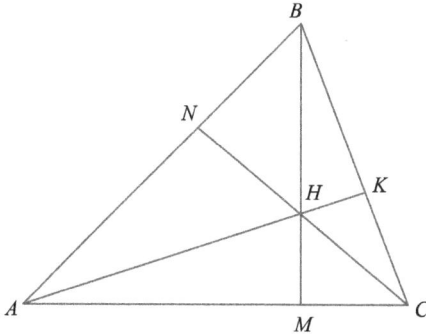

Figure 1

Since ancient Greek times, the orthocenter of a triangle is known as one of the most intriguing of its associated points.

First of all, we have to prove that such a point exists and that the three altitudes pass through it. According to some historians the first such proof is attributable to Proclus (412–485 AD). Various proofs were discovered through the years. We will present Leonhard Euler's proof — very insightful and easy to follow.

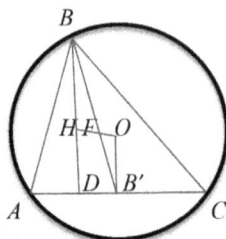

Figure 2

In Figure 2, triangle ABC is inscribed in a circle with center O. BB' is its median and BD is its altitude, point F is the centroid (the point of intersection of the medians). Let's consider the point H, such that F is located between H and O and $HF = 2FO$. Our goal is to prove that H is the orthocenter of the triangle ABC.

Observe that by the construction, $HF = 2OF$ and $BF = 2B'F$ by the property of the centroid to divide the medians in the ratio 2:1. Also, angles HFB and OFB' are congruent as vertical angles. It follows that triangles HFB and OFB' are similar, which implies that the pairs of other respective angles in these triangles are congruent. Thus BH is parallel to OB' because if a pair of alternate interior angles formed by a transversal of two straight lines are congruent, the lines are parallel. Next, recall that the circumcenter of a triangle O is the point where three perpendicular bisectors from the sides of a triangle intersect. Hence, since OB' is perpendicular to AC, BH being parallel to OB', must be also perpendicular to AC. In other words, H lies on the altitude BD. In a similar fashion, we can prove that H lies on each of the other two altitudes as well. Therefore, the altitudes do meet at one point — the orthocenter of the triangle, and this point indeed exists.

Did you notice the remarkable fact established during the proof: the circumcenter, orthocenter, and centroid of a triangle lie on the same straight line, the Euler line? It was named after the prominent mathematician Leonhard Euler (1707–1783), who was the first to discover this property and showed that the centroid divides the distance from the orthocenter to the circumcenter in the ratio 2:1. We will come across the Euler line one more time in one of the subsequent chapters, as an unexpected by-passing effect of reviewing various solutions to a non-standard construction problem. Wouldn't it be challenging to get a different view at that property?

The orthocenter of a triangle may often come to the rescue in many difficult problems, including constructions with so-called restricted conditions. Sometimes the key to a solution is to simply apply the fact of the orthocenter's existence in a given triangle. Here are a few examples.

Problem 1 (Jacob Steiner's problem). Given a circle with a diameter AB and point M outside the circle, construct the perpendicular to AB passing through point M using only a one-sided unmarked ruler. (You can only draw straight lines using that ruler.)

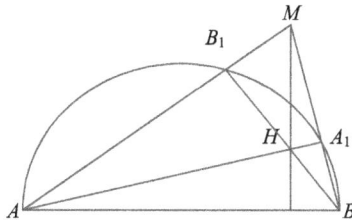

Solution. A straight line drawn from A to M intersects the circumference at point B_1 and the straight line from B to M at point A_1. $\angle AB_1B = \angle AA_1B = 90°$, since both angles are inscribed in a semicircle. Then BB_1 and AA_1 are altitudes in triangle AMB. Let them intersect at the point H, which is the orthocenter of triangle AMB. The third altitude must pass through H. If we draw MH, then it must be perpendicular to AB. The desired result is achieved.

Problem 2. A straight line m is given on a plane. Draw an arbitrary straight line perpendicular to m using only a marked one-sided ruler.

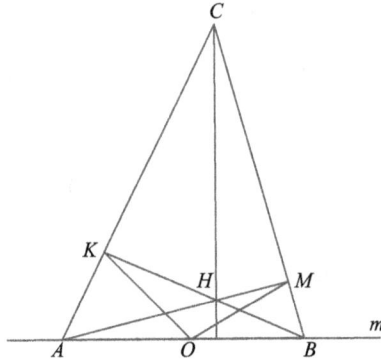

Solution. We choose any point O on m and draw equal segments $OA = OB = 1$, where points A and B lie on m. Next, we draw any equal segments $OK = OM = 1$, where K and M lie on the same side of m. Draw lines AK and BM intersecting at C. By the construction, $OA = OB = OK = OM$, and we may conclude that points A, K, M, and B must lie on the circumference of the circle with center O and radius 1. Since AB is the diameter of that circle, inscribed angles AKB and AMB both equal to 90°. Therefore, BK and AM are altitudes in triangle ACB. The point of their intersection H is the orthocenter of triangle ABC. The final step is to draw the line through C and H. The line CH is the desired line perpendicular to line m.

Problem 3. Let there be a right triangle ACB ($\angle C = 90°$). CD is an altitude of triangle ACB. M and N are the midpoints of CD and BD, respectively. Prove that $AM \perp CN$.

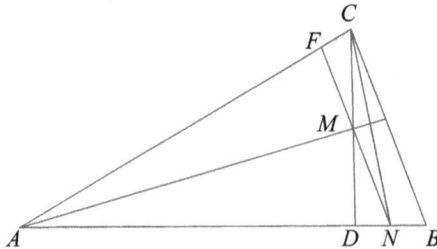

Proof. It's given that $CM = MD$ and $DN = NB$, thus in triangle CDB, MN is the midline and $MN \parallel CB$. Also, $BC \perp AC$ by the conditions of the problem. Therefore, $MN \perp AC$. Let MN intersect AC at the point F. NF and CD are altitudes in triangle CAN. The third altitude contains their point of intersection M. So, $AM \perp CN$, which was to be proved.

Now let's turn to some important properties of the orthocenter of a triangle.

Theorem 1. *Points symmetrical to the orthocenter of a triangle with respect to the midpoints of its sides lie on its circumcircle.*

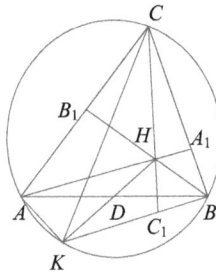

Figure 3

Proof. In the triangle ABC at Figure 3, AA_1, BB_1, and CC_1 are its altitudes, H is its orthocenter, and D is the midpoint of AB. K is symmetrical to H with respect to D, and therefore, $DK = DH$.

We need to prove that point K lies on the circumcircle of the triangle ABC.

First, we will prove that the sum of angles ACB and AHB is $180°$ (if angle A is less than $90°$ and angle B is less than $90°$). In the quadrilateral HB_1CA_1, $\angle A_1 = \angle B_1 = 90°$, because BB_1 and AA_1 are altitudes in triangle ABC. Hence, $\angle B_1CA_1 + \angle B_1HA_1 = 180°$.

Also, $\angle B_1HA_1 = \angle AHB$ because they are vertical angles. Hence,

$$\angle ACB + \angle AHB = \angle B_1CA_1 + \angle B_1HA_1 = 180°. \tag{1}$$

Because $AHBK$ is a parallelogram (diagonals are split in half by their point of intersection D), its opposite angles are equal,

$\angle AKB = \angle AHB$. After substituting this into (1), we get that the opposite angles in the quadrilateral $ACBK$ supplement each other: $\angle AKB + \angle ACB = 180°$. By the theorem about quadrilaterals inscribed in a circle (cyclic quadrilaterals), $ACBK$ is a cyclic quadrilateral (you may find the detailed proof in Chapter 5). As vertices of this cyclic quadrilateral, points A, C, B, and K lie on that circle, and we are done.

I want to emphasize the important fact established during the proof of Theorem 1:

The angle formed at the orthocenter by the altitudes is the supplement of the angle at the vertex ($\angle AHB + \angle ACB = 180°$, see Figure 3).

Theorem 2. *The images of the orthocenter of a triangle by reflection in its sides lie on the circumcircle of the triangle.*

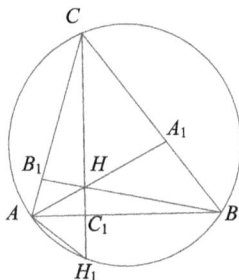

Figure 4

Proof. Let AA_1 and CC_1 be altitudes in triangle ABC, and H the orthocenter. In triangle ABA_1, $\angle A_1 AB = 90° - \angle B$; in triangle CBC_1, $\angle C_1 CB = 90° - \angle B$.

It follows that $\angle A_1 AB = \angle C_1 CB$.

Extend CC_1 to its intersection with the circumcircle of triangle ABC at point H_1. Our goal is to prove that H_1 is an image of H by reflection in AB.

$\angle H_1 AB = \angle H_1 CB$ because both angles are inscribed in the same arc $H_1 B$ of the circle. It was already proved that $\angle A_1 AB = \angle C_1 CB$; so $\angle H_1 AB = \angle A_1 AB$ or $\angle HAC_1 = \angle H_1 AC_1$. We get two right triangles $AC_1 H$ and $AC_1 H_1$ with the common side AC_1 and congruent acute angles at vertex A. Therefore, the triangles are congruent and respectively, $HC_1 = H_1C_1$. By definition, points H and H_1 are symmetrical with respect to side AB. We see that points H_1, A, C, and B lie on one circle, the circumcircle of triangle ABC, which was to be proved.

An immediate consequence of the theorem is:

Three points where the extended altitudes meet the circumcircle form a triangle similar to the orthic triangle.

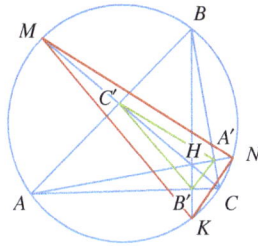

Proof. Let the extended altitudes AA', BB', and CC' of the triangle ABC intersect the circumcircle at points N, K, and M, respectively. We need to prove that triangles $A'B'C'$ and NKM are similar.

According to Theorem 2, points N, K, and M are the images of the orthocenter H by reflection in its sides. Then $HA' = A'N$, $HB' = B'K$, $HC' = C'M$ and we see that $A'B'$, $B'C'$, and $C'A'$ are the midlines in the triangles HNK, HKM, and HMN respectively, which implies that $A'B' = \frac{1}{2}KN$ and $A'B' \parallel KN$; $B'C' = \frac{1}{2}MK$ and $B'C' \parallel MK$; $A'C' = \frac{1}{2}MN$ and $A'C' \parallel MN$. Hence, all the respective sides of our two triangles are parallel and their respective ratios equal $\frac{1}{2}$, which means that triangles $A'B'C'$ and NKM are similar with the ratio $\frac{1}{2}$, and the proof is completed.

We leave to readers to prove another interesting property derived also as a corollary of the theorem:

In any triangle, the products of the lengths of the segments that the orthocenter divides an altitude into is the same for all altitudes: $AH \cdot HA_1 = BH \cdot HB_1 = CH \cdot HC_1$ (see Figure 4).

Problem 4. Given three points K, M, and N as the points of intersection of a triangle's three altitudes with its circumcircle, construct the triangle.

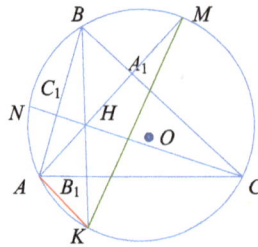

Solution. This problem can easily be solved by applying Theorem 2.

Assume the problem is solved and we see in figure above, the triangle ABC with its circumcircle and its three altitudes AA_1, BB_1, and CC_1, extensions of which intersect the circumcircle at points M, K, and N respectively. Let's analyze the problem and find a plan to construct the triangle ABC by having only the three given points M, K, and N.

As it was proved in Theorem 2, point K is an image of the ortho-center H by reflection in AC. Then $\angle HAC = \angle KAC$ or, equivalently, $\angle MAC = \angle KAC$. The equality of the inscribed angles implies that C divides the arc MK in half and is equidistant from K and M. Therefore, in order to locate the point C on the circumcircle, we would need to draw the perpendicular bisector to segment KM to its intersection with the circumcircle. In the same fashion, we would get points A and B.

The plan for the construction is the following.

First, draw three perpendicular bisectors to the segments connecting the pairs of three given points K, M, and N and get the point of their intersection O. Then draw the circle with O as the center and radius $r = OK = OM = ON$. Finally, extend the perpendicular bisectors of segments MN, MK, and KN to their intersections with the circle at points B, C, and A, the vertices of the triangle ABC. ABC is the desired triangle. We leave rigorous proofs of our assertions for the readers to complete as well as to investigate how many solutions the problem has.

Let's now take the next step and talk about some surprising relationships between altitudes and other elements in a triangle, that can be established by using the formula for the area of a triangle $S = \frac{1}{2}ah$ (one-half the product of the triangle's base a and the altitude h dropped to that base). We will take advantage of the method below for deriving nice and easy solutions to several interesting problems that are rather difficult to solve directly. Make a note of it as it will be revisited a few times in the subsequent chapters.

Problem 5. Prove that the area of a triangle may be calculated by the formula $S = \frac{1}{2}Pr$, if P is the triangle's perimeter and r is its inradius.

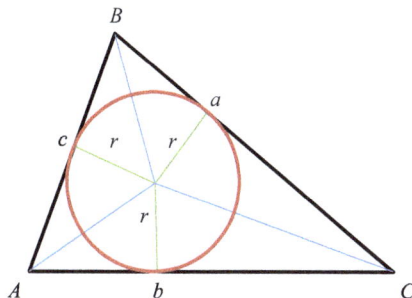

Proof. By connecting the vertices of the triangle to the center of its incircle, we partitioned triangle ABC into three triangles. In each the inradius is the altitude dropped to the respective side, therefore the areas of the triangles are:

$$S_1 = \frac{1}{2}ar,$$

$$S_2 = \frac{1}{2}br, \text{ and}$$

$$S_3 = \frac{1}{2}cr.$$

Adding, we get the area of the triangle ABC:

$$S = \frac{1}{2}(a + b + c)r = \frac{1}{2}Pr.$$

Problem 6 (Inradius Theorem). Prove that for a triangle with altitudes h_a, h_b, h_c, and inradius r:

$$\frac{1}{h_a} + \frac{1}{h_b} + \frac{1}{h_c} = \frac{1}{r}.$$

Proof. The area of the triangle can be found as $S = \frac{1}{2}ah_a$, $S = \frac{1}{2}bh_b$, or $S = \frac{1}{2}ch_c$. Then

$$\frac{1}{h_a} = \frac{a}{2S} \tag{1}$$

$$\frac{1}{h_b} = \frac{b}{2S} \tag{2}$$

$$\frac{1}{h_c} = \frac{c}{2S} \tag{3}$$

Adding (1), (2), and (3), we get

$$\frac{1}{h_a} + \frac{1}{h_b} + \frac{1}{h_c} = \frac{a+b+c}{2S} = \frac{P}{2S}, \tag{4}$$

where P is the triangle's perimeter.

As it was proved in Problem 5, the area of the triangle may be found as

$$S = \frac{1}{2}Pr. \tag{5}$$

Comparing (4) and (5) it follows that $\frac{1}{h_a} + \frac{1}{h_b} + \frac{1}{h_c} = \frac{1}{r}$, and we are done.

Problem 7. Prove that for a triangle with altitudes h_a, h_b, h_c and inradius r,

$$h_a + h_b + h_c \geq 9r.$$

Proof. It's obvious that for any two positive numbers a_1 and a_2, $\frac{a_1}{a_2} + \frac{a_2}{a_1} \geq 2$.

Indeed, if $a_1 > 0$ and $a_2 > 0$, then $\frac{a_1}{a_2} + \frac{a_2}{a_1} - 2 = \frac{a_1^2 + a_2^2 - 2a_1a_2}{a_1a_2} = \frac{(a_1-a_2)^2}{a_1a_2} \geq 0$.

By applying this, it is easy to prove that $(a_1 + a_2 + a_3)\left(\frac{1}{a_1} + \frac{1}{a_2} + \frac{1}{a_3}\right) \geq 9$.

Consider $(a_1 + a_2 + a_3)\left(\frac{1}{a_1} + \frac{1}{a_2} + \frac{1}{a_3}\right) = 1 + 1 + 1 + \left(\frac{a_1}{a_2} + \frac{a_2}{a_1}\right) + \left(\frac{a_1}{a_3} + \frac{a_3}{a_1}\right) + \left(\frac{a_3}{a_2} + \frac{a_2}{a_3}\right) \geq 3 + 2 + 2 + 2 = 9$.

In Problem 6, we proved that $\frac{1}{h_a} + \frac{1}{h_b} + \frac{1}{h_c} = \frac{1}{r}$. Therefore,

$$(h_a + h_b + h_c)\left(\frac{1}{h_a} + \frac{1}{h_b} + \frac{1}{h_c}\right) = (h_a + h_b + h_c)\cdot\frac{1}{r} \geq 9.$$

Multiplying both sides of the last inequality by a positive number r, we get $h_a + h_b + h_c \geq 9r$, as we wanted to prove.

What would you say about a triangle in which three altitudes satisfy $h_a + h_b + h_c = 9r$? It's not a trivial question. How much time would it take to get an answer if you face this problem at the Math Olympiad?

Curiously, the answer becomes obvious by using the previous problem. It was proved that for any triangle, $h_a + h_b + h_c \geq 9r$. Then the equality $h_a + h_b + h_c = 9r$ will be possible only when $h_a = h_b = h_c$, which means that such a triangle has to be a regular triangle (triangle with all three sides of equal length).

Problem 8 (Circumradius Theorem). Prove that for a triangle with the altitude from one side h_a, the other two sides as b and c, and the triangle's circumradius R, $h_a = \frac{bc}{2R}$.

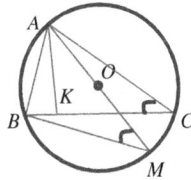

Proof. In the triangle ABC denote $h_a = AK$, $b = AC$, $c = AB$, $a = BC$, $R = AO = OM$.

AM is the diameter of the circumcircle of the triangle ABC. The inscribed angle ABM is subtended by AM, thus it must be a right angle. $\angle AMB = \angle ACB$ because they are inscribed angles subtended by the same chord AB. Then in the right triangle ABM,

$$c = AB = AM \cdot \sin\angle AMB = 2R\cdot\sin\angle AMB, \text{ from which } \sin\angle AMB = \frac{c}{2R}. \quad (*)$$

The area of the triangle ABC is
$S = \frac{1}{2}ba \cdot \sin\angle ACB = \frac{1}{2}ba \cdot \sin\angle AMB$ or $S = \frac{1}{2}ah_a$. Comparing the formulas and substituting the value of $\sin\angle AMB$ from (*), it's easy

to get that $h_a = \frac{bc}{2R}$. Obviously, similar equalities will hold for the other sides and respective altitudes of a triangle as well.

As you have just seen, by linking problems together and applying concepts and formulas for the altitude and area of a triangle, you may not only get elegant and relatively simple solutions, but you may also explore the unexpected relationships and connections among a triangle's linear elements: sides, altitudes, inradius, and circumradius.

We invite you to prove another two properties on your own:

Given a triangle ABC with altitudes AD, CF, and BE. Point H is its orthocenter. Prove that

$$HD/AD + HE/BE + HF/CF = 1.$$

$$AH/AD + BH/BE + CH/CF = 2.$$

Finally, let us talk about a very interesting and unique mathematical phenomenon — the **Nine-Point Circle**, which is usually referred to as *Euler's Circle*. Leonhard Euler was once thought to be its earliest discoverer. Called by Pierre Simon Laplace as master of all mathematicians, the prominent Swiss scholar made significant contributions to virtually every field of mathematics. At the beginning of this chapter, we came across the *Euler line* and its properties. Ironically, *Euler's circle* is perhaps not an appropriate name because there is no evidence in Euler's writings that he was the first to discover it. In 1765 he mentioned six points of a triangle lying on one circle — the midpoints of the sides and the feet of the altitudes. Three more points on the same circle are the midpoints of segments from the vertices of a triangle to its orthocenter. This was noticed by Charles Julien Brianchon (1783–1864) in collaboration with Jean Victor Poncelet (1788–1867) in 1820.

There have been several independent discoverers of the nine-point circle. German mathematician Karl Wilhelm Feuerbach (1800–1834) was the first to publish the proof that the nine-point circle is tangent to the incircles and excircles of a triangle, but he was talking about the six-point circle. Orly Terquem, one of the editors of the *Nouvells Annales*, was the first to designate the circle the **nine-point circle.** That is why it bears the other names and is known also as *Feuerbach's circle, Terquem's circle, medioscribed circle, n-point circle,* and *the* *mid circle* (to name a few). For further reading on the history of the subject, you may see "History of The Nine Point Circle" by J.S. MacKay, which was published in the *Proceedings of the Edinburgh Mathematical Society* in 1892.

The nine-point circle for a triangle passes through the three feet of the altitudes, the three midpoints of the sides, and the three midpoints of the segments connecting the vertices to the orthocenter. The center of the circle is the midpoint of the segment with endpoints at the orthocenter and the circumcenter.

In figure below, ABC is the given triangle; M, L, and K are the feet of the altitudes on the sides; F, E, and D are the midpoints of the sides; H is the orthocenter; P, Q, and R are the midpoints of segments AH, BH, and CH, respectively.

Our goal is to prove that all nine points F, M, P, L, E, R, K, D, and Q lie on the same circle.

Proof.

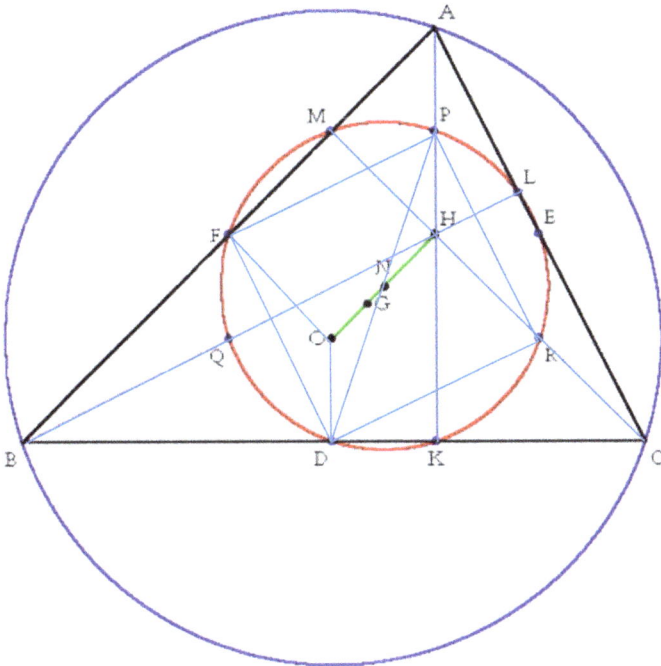

Because FD is the midline in triangle ABC, $FD \parallel AC$ and $FD = \frac{1}{2}AC$ by the Midline Theorem. PR is the midline in triangle AHC, thus $PR \parallel AC$ and $PR = \frac{1}{2}AC$. By transitivity, we conclude that $FD \parallel PR$ and $FD = RP$. Likewise, we get that $FP \parallel DR$ and $FP = DR$. It follows that $FPRD$ is a parallelogram. It's not hard to prove that it is a rectangle. Indeed, since $BL \perp AC$ and $FP \parallel BL$, $PR \parallel AC$, then $FP \perp PR$. We conclude that the points F, P, R, and D being the vertices of a rectangle lie on the circle with the center at the point of intersection of the diagonals and radius equals one-half the diagonal. If we denote by N the midpoint of DP, then N is the center of the circle and its radius is $NP = ND$. Now let's focus on the right triangle PKD (P lies on the altitude AK dropped to BC, so $PK \perp DK$). N is the midpoint of its hypotenuse, therefore it is the center of the circumcircle of the triangle PKD. It is the same circle we just proved is circumscribed about the rectangle $FPRD$ (DP is the diameter in both cases). Therefore, K lies on the same circle as points F, P, R, and D. In a similar manner, following the same procedures, we obtain that the other four points Q, M, L, and E also lie on the same circle concluding our proof of the existence of the nine-point circle.

You may have noticed another interesting relationship emanating from the above proof — the nine-point center is collinear with the orthocenter, centroid, and circumcenter in a triangle, being the fourth important point lying on Euler's line. The four points will coincide only in an equilateral triangle, but in any other triangle they are distinct from each other.

In the next chapter, we will continue the study of the properties of the orthocenter and the altitudes of a triangle. We will concentrate on properties of the orthic triangle.

In this chapter, we covered the cases where the orthocenter was lying inside its triangle. In fact, the orthocenter lies inside the triangle if and only if it is acute triangle. In a right triangle, the orthocenter coincides with the vertex of the right angle. We leave the properties of the orthocenter of obtuse triangles and a few problems for readers' exploration.

Problem 9. Let A be an inaccessible vertex of triangle ABC. Draw the line perpendicular to BC passing through A.

Problem 10. Given a circle, A lies on the circle and H is inside the circle. Construct points B and C on the circle such that H is the orthocenter of triangle ABC.

Problem 11. Let H be the orthocenter of triangle ABC. Prove that the radii of the circumcircles of triangles ABC, AHC, AHB, and BHC are equal.

Chapter 3

The Orthic Triangle and Its Properties

Of all triangles inscribed in a triangle, the most interesting properties are in its orthic triangle. Even though many of the properties seen later in this chapter are classic examples of relationships in a triangle, they are not often considered in high school classrooms. Their study is, however, a positive experience for students, one in which they may explore the world of plane geometry, make conjectures based on patterns discovered, and deduce the geometrical concepts. In this chapter, we restrict our attention to those properties that are most useful and helpful for solving problems.

Our goal is to give the reader not only facts about the orthic triangle, but insight into methods of geometric proof. Each property is not to be looked at individually but as part of a whole. We will show how one proof generalizes another and how the proofs can be linked intuitively.

First, we recall the definition of the orthic triangle.

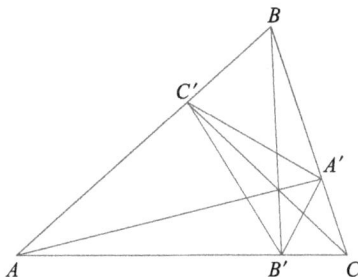

Draw the altitudes of the triangle ABC, AA', BB', and CC'. The points A', B', and C' are naturally called the feet of the altitudes. Joining them in pairs we obtain the triangle $A'B'C'$ which is called the orthic triangle of $\triangle ABC$. When ABC is a right triangle, the orthic triangle degenerates into a line segment — an altitude to the hypotenuse.

To begin, let's consider three introductory theorems.

Theorem 1. *Let AA' and CC' be altitudes of the triangle ABC. Then triangles ABC and $A'BC'$ are similar.*

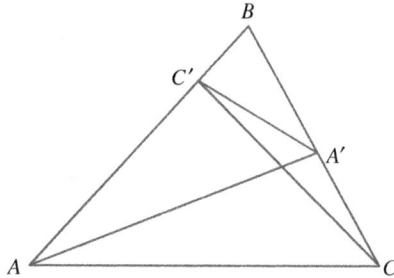

Proof.   In triangle $AA'B$, $\angle A' = 90°$, $BA' = BA \cos B$, and in triangle $CC'B$, $\angle C' = 90°$, $BC' = BC \cos B$.

We see that $BA':BC' = BA:BC$ and $\angle B$ is the common angle in the triangles ABC and $A'BC'$. Therefore, these triangles are similar with the ratio $k = \cos B$, which was what had to be proved.

We have provided the proof for an acute triangle. For an obtuse triangle the proof would be similar. In the case of a right triangle, the altitude drawn to the hypotenuse, in which the orthic triangle degenerates, divides the triangle into two similar right triangles, each of which is similar to the given right triangle.

From the theorem, it follows immediately that $\angle BAC = \angle BA'C'$ and $\angle BCA = \angle BC'A'$, or expressing it in words:

Corollary. *The angles of a triangle cut by the sides of the orthic triangle of the initial triangle are equal to its corresponding angles.*

This important fact may be used as the connecting link between the proofs of almost all of the following properties, and also as the base building block in the development of solutions to many problems that involve the orthic triangle.

Theorem 2. *The altitudes of an acute triangle ABC bisect the angles of its orthic triangle $A'B'C'$.*

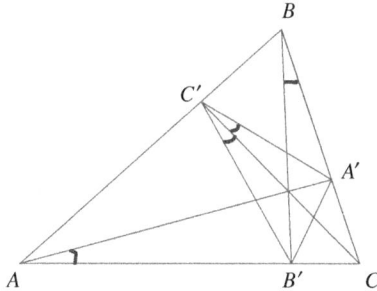

Proof. At first glance, the discovery of a proof of the theorem may seem difficult, and for some readers even confusing (how is it possible to link the altitudes of the triangle with the bisectors of the angles of its orthic triangle?).

It turns out, however, to be fairly straightforward when using the corollary from Theorem 1. In order to prove the theorem, we merely need to observe that $\angle AC'B' = \angle BC'A'$, because from the corollary from Theorem 1, each of these angles is equal to $\angle ACB$. And, noting that $\angle AC'C = \angle BC'C = 90°$, we get that $\angle B'C'C = 90° - \angle AC'B' = 90° - \angle BC'A' = \angle A'C'C$. Or, expressing it in words, $C'C$ is the bisector of the angle $A'C'B'$.

In the same way, we get that $A'A$ and $B'B$ are the bisectors of the other angles.

Let's prove the immediate important corollary from the theorem: $\angle A'C'B' = 180° - 2\angle ACB$.

First, note that $\angle A'AC = \angle B'BC = \angle B'C'C = \angle CC'A'$.

Indeed, angles $A'AC$ and $B'BC$ are equal as respective complements of angle C in the right triangles $AA'C$ and $BB'C$. Also, since the right triangles $AC'C$ and $AA'C$ have the common hypotenuse AC, all four points A, C', A', and C lie on the single circle with diameter AC. Then angles $A'AC$ and $CC'A'$ are equal as inscribed angles that subtend the same arc $A'C$.

Similarly, $\angle B'BC = \angle B'C'C$. To finish the proof, we need to express angle C from triangle ACA' (or you can use triangle $B'CB$): $\angle ACB = \angle ACA' = 90° - \angle A'AC = 90° - \angle B'C'C = 90° - \frac{1}{2}\angle A'C'B'$. From the last equality, it follows that $\angle A'C'B' = 180° - 2\angle ACB$.

The proof of Theorem 2, as well as all the proofs in this chapter, is of course, not unique. It would be a useful exercise for readers to find other proofs. For example, consider the cyclic quadrilateral formed by the sides of the triangle ABC and segments of the altitudes between the orthocenter and the points A', B', and C'. We will use these cyclic quadrilaterals in several problems which follow.

Theorem 3. *The line segment joining a vertex of a triangle to the center of its circumcircle is perpendicular to the corresponding side of the orthic triangle.*

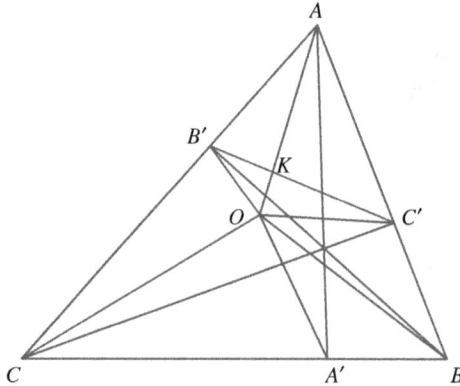

Proof. If O is the center of the circumcircle of the triangle ABC, then $AO = BO = CO = R$, where R is the radius of the circumcircle. Consider the cyclic quadrilaterals $OB'AC'$, $OC'BA'$, and $OA'CB'$. Their diagonals OA, OB, and OC form with the sides of the triangle ABC pairs of equal angles. Denote them by

$$\angle 1 = \angle OBA = \angle OAB,$$

$$\angle 2 = \angle OBC = \angle OCB,$$

$$\angle 3 = \angle OAC = \angle OCA.$$

In triangle ABC, $\angle A + \angle B + \angle C = 180°$, hence
$\angle 1 + \angle 2 + \angle 3 = \frac{180°}{2} = 90°$.

Then $\angle 1 + \angle ACB = 90°$ because $\angle 2 + \angle 3 = \angle ACB$. From
Theorem 1, we know that triangles ABC and $AB'C'$ are similar, thus
$\angle ACB = \angle AC'B'$. If we denote by K the point of intersection of the
lines OA and $B'C'$, then in triangle AKC', $\angle KC'A + \angle KAC' = 90°$,
i.e., $AK \perp B'C'$, which is what we wished to prove.

The problems that follow are often found in textbooks in special sections for difficult problems, and some of them have appeared
in mathematical contests. The solutions of all these problems may
be significantly simplified by using the techniques outlined earlier.
Theorems 1, 2, 3 and the corollary of Theorem 1 are the basic building blocks in the development of solutions.

Problem 1. AA', BB', and CC' are the altitudes of the triangle
ABC.

Prove that $AB' \cdot BC' \cdot CA' = AC' \cdot BA' \cdot CB' = B'C' \cdot C'A' \cdot A'B'$.

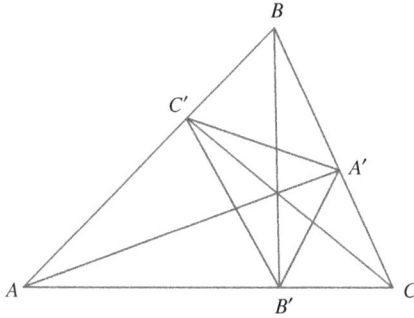

Proof. The easiest solution can be obtained by the direct use of
Theorem 1. Since BAC and $B'AC'$ are similar triangles with a ratio
$k = \cos A$, then each side of the triangle $B'AC'$ may be expressed as
the product of the corresponding side of the triangle ABC by $\cos A$:

$$AB' = AB \cos A,$$
$$AC' = AC \cos A,$$
$$B'C' = BC \cos A.$$

Analogously, from the other two pairs of similar triangles $A'BC'$ and ABC, $A'CB'$ and ABC:

$$BC' = BC \cos B,$$
$$BA' = AB \cos B,$$
$$C'A' = AC \cos B.$$
$$CA' = AC \cos C,$$
$$CB' = BC \cos C,$$
$$A'B' = AB \cos C.$$

By multiplying respectively the first, second, and third of the equalities, we obtain

$$AB' \cdot BC' \cdot CA' = AC' \cdot BA' \cdot CB' = B'C' \cdot C'A' \cdot A'B' = AB \cdot BC \cdot AC \cos A \cos B \cos C,$$

which is the desired equality.

In Problems 2–4, it is worthwhile to use the powerful tool of trigonometry, especially when working with right triangles. Trigonometric formulas with suitable manipulations can provide the link between the earlier theorems and the following problems. I would hope that Problems 2–4 will be a challenge to the reader and will be tried before continuing on with this chapter.

Problem 2. Let S be the area of an acute-angled triangle ABC, and S' be the area of its orthic triangle $A'B'C'$. Prove that $\frac{S'}{S} = 2 \cos A \cos B \cos C$.

You might want to get back to this problem again after reading the chapter "The Theorem of Ratios of Areas of the Similar Polygons" and applying the methods outlined there.

Problem 3. Prove that the radius of the circumcircle of a triangle is twice the radius of the circumcircle of its orthic triangle.

Problem 4. ABC is an acute-angled triangle with sides a, b, and c in the usual notation. Prove that there exists a triangle with sides $a \cos A$, $b \cos B$, $c \cos C$.

Solutions to Problems 5 and 6 are not closely related to each other, but each of them is based on the results derived earlier. Both problems belong to "orthic triangle" type of problem. In Problem 5 recognizing the altitudes of the given triangle as the bisectors of its orthic triangle leads to the desired result. In Problem 6, the solution gained ground through cyclic quadrilaterals and results obtained in Theorem 3.

An important technique that will be used in Problem 6 is that of calculating the area of an *orthic quadrilateral* (a quadrilateral with perpendicular diagonals) as half the product of its diagonals.

Problem 5. In an acute triangle ABC, AA', BB', and CC' are its altitudes. L is a point on AC such that $C'L \parallel BC$. Prove that $\angle CC'L = \frac{1}{2}\angle C'B'A'$.

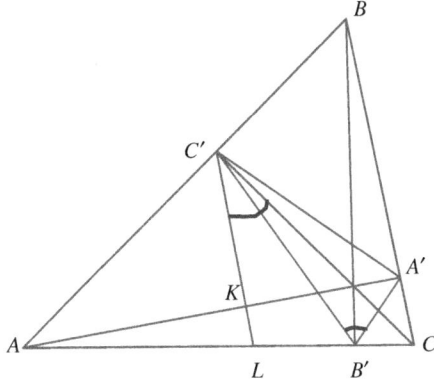

Proof. By Theorem 2, $B'B$, $A'A$, and $C'C$ bisect the corresponding angles of the orthic triangle $A'B'C'$.

Denote $\angle 1 = \angle C'B'B = \angle A'B'B$, $\angle 2 = \angle B'C'C = \angle A'C'C$, $\angle 3 = \angle C'A'A = \angle B'A'A$, and for convenience, let's denote $\angle x = \angle LC'B'$. Then

$$\angle 1 + \angle 2 + \angle 3 = \frac{180°}{2} = 90°. \tag{1}$$

$C'L \parallel BC$ and $AA' \perp BC$, thus $AA' \perp C'L$.

If K is the point of intersection of AA' and $C'L$, then $\angle C'KA' = 90°$. In the right triangle $C'KA'$, angles $KC'A'$ and $KA'C'$ are complementary.

Since $\angle KA'C' = \angle 3$ and $\angle KC'A' = \angle x + \angle 2 + \angle 2$, we get

$$\angle x + \angle 2 + \angle 2 + \angle 3 = 90°. \tag{2}$$

Comparing (1) and (2) gives $\angle x + \angle 2 = \angle 1$, from which $\angle x = \angle 1 - \angle 2$.

$\angle LC'C = \angle x + \angle B'C'C = \angle x + \angle 2 = \angle 1 - \angle 2 + \angle 2 = \angle 1$.

It follows that indeed, the desired angle $LC'C$ is half of angle $C'B'A'$.

When $C'L$ lies between $C'B'$ and $C'C$, the solution will be similar. It's interesting to note that $C'L$ would lie between $C'B'$ and $C'C$ when the angle ABC is greater than the angle ACB, and points B' and L would coincide when the angles are equal. As an exercise, readers can prove it by using, for instance, the corollary from Theorem 1.

Problem 6. Let R be the radius of the circumcircle of the acute triangle ABC.

AA', BB', and CC' are the altitudes of the triangle ABC, p is the semi-perimeter of the orthic triangle $A'B'C'$, and S is the area of the triangle ABC.

Prove that $S = pR$.

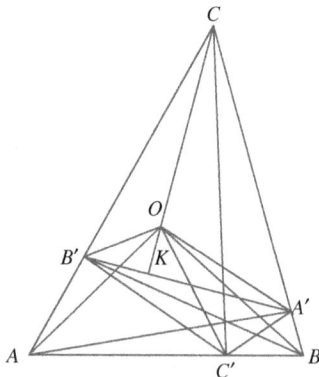

Proof. Denote by S_1 the area of the quadrilateral $OB'AC'$, S_2 the area of the quadrilateral $OC'BA'$, and S_3 the area of the quadrilateral $A'CB'O$.

From Theorem 3, we get that the diagonals of the quadrilaterals are perpendicular:

$OA \perp B'C'$ and $OB \perp A'C'$. A quadrilateral with perpendicular diagonals is called orthic quadrilateral, and it's not hard to prove (the proof of this is left to the reader) that its area can be calculated as half the product of the diagonals. Therefore,

$$S_1 = \frac{1}{2} OA \cdot B'C' = \frac{1}{2} R \cdot B'C'; \tag{1}$$

$$S_2 = \frac{1}{2} OB \cdot A'C' = \frac{1}{2} R \cdot A'C'. \tag{2}$$

Let CO extended meet $B'A'$ at K. Then $CK \perp A'B'$.
When O lies outside the triangle $A'B'C'$ (as it is in figure above), S_3 is the difference between the areas of the triangles $A'CB'$ and $A'OB'$:

$$S_3 = \frac{1}{2} CK \cdot A'B' - \frac{1}{2} OK \cdot A'B' = \frac{1}{2}(CK - OK)A'B' = \frac{1}{2} R \cdot A'B'. \tag{3}$$

When O lies inside the triangle $A'B'C'$, S_3 is the sum of the areas of the triangles $A'CB'$ and $A'OB'$:

$$S_3 = \frac{1}{2} CK \cdot A'B' + \frac{1}{2} OK \cdot A'B' = \frac{1}{2}(CK + OK)A'B' = \frac{1}{2} CO \cdot A'B' = \frac{1}{2} R \cdot A'B',$$

which is the same result as obtained above. The diagram though would be a little different (it is not provided in this solution and we encourage readers to explore that case on their own).

By noting that $S = S_1 + S_2 + S_3$ and adding (1), (2), and (3), we find that $S = \frac{1}{2}(B'C' + A'C' + A'B')R = pR$, which is the desired result.

Finally, we are ready to turn to one of the most remarkable properties of the orthic triangle, which is known as **Fagnano's Problem:**

Of all triangles inscribed in an acute triangle, the orthic triangle has the smallest perimeter.

This problem was proposed in 1775 by Giovani Francesco Fagnano dei Toschi (1715–1797), who solved it by means of differential calculus. There are several other solutions, perhaps the best known being that of Hermann Schwarz.

Karl Hermann Amandus Schwarz (1843–1921) introduced using reflections to solve the problem, a technique which is useful in many orthic triangle problems. His method was much simplified by his student Lipót Fejér (1880–1959). This solution can be found in the second edition of *Introduction to Geometry* by H.S.M. Coxeter (pp. 20–21).

When I first came across this interesting problem in math literature, I was fascinated by the beauty and elegance of Schwarz's solution and asked myself if it was possible to find the proof by applying the least possible reflections. I managed to get the result by employing only two reflections. In comparison with Fejér's proof, it may seem more complicated and less elegant, but it is worthwhile discussing as it follows in the sequence of the orthic triangle properties discussed in this chapter.

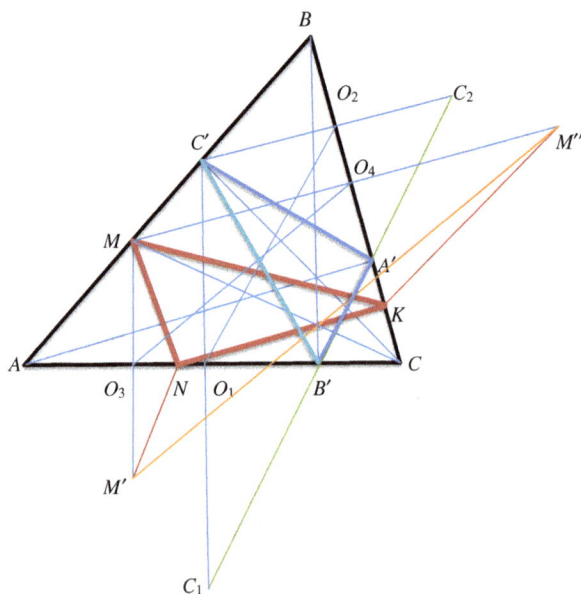

Proof. In triangle ABC, we draw $AA' \perp BC$, $BB' \perp AC$, and $CC' \perp AB$. Connecting A', B', and C' we obtain the orthic triangle of ABC.

Our goal is to compare the perimeter of the triangle $A'B'C'$ with the perimeter of a random triangle MNK inscribed in $\triangle ABC$, and show that the perimeter of the orthic triangle is less than the perimeter of a randomly inscribed triangle.

The main idea behind this proof was to compare the length of the segment C_1C_2(points C_1 and C_2 are the images of C' by reflection in AC and BC, respectively), which equals the perimeter of the orthic triangle $A'B'C'$, with the length of the segment $M'M''$ joining the images M' and M'' of the vertex M of an arbitrary triangle MNK inscribed in the triangle ABC by reflection in AC and BC, respectively (M, N, and K lie on AB, AC, and BC, respectively).

The perimeter of the triangle MNK equals the length of the broken line $M'NKM''$. The length of any broken line is always greater than the length of the segment joining its endpoints. Thus, if we manage to prove that $M'M'' > C_1C_2$, we will be able to achieve the desired result. When M', N, K, and M'' lie on the same straight line, the perimeter of the triangle MNK is equal to the length of $M'M''$. So, in either case, we must prove that $M'M'' > C_1C_2$.

First, let's show that, as claimed above, the perimeter of the orthic triangle equals the length of the segment C_1C_2.

By the corollary to Theorem 1, $\angle C'B'A = \angle A'B'C = \angle ABC$. Hence, observing the angles formed at point B', we have $\angle C'B'A' = 180° - 2\angle ABC$.

Also, by the properties of reflection, $\angle C'B'A = \angle C_1B'A$ and thus $\angle C_1B'A + \angle AB'C' + \angle C'B'A' = \angle B + \angle B + 180° - 2\angle B = 180°$, which proves that C_1, B', and A' lie on the same straight line. Similarly, the point C_2 lies on this line as well.

$C'B' = C_1B'$ and $A'C' = A'C_2$ by the properties of reflection. Then the perimeter of the triangle $A'B'C'$ is $C'B' + A'B' + A'C' = C_1B' + A'B' + A'C_2 = C_1C_2$, which is what we set out to show above.

Next, denote by O_1, O_2, O_3, and O_4 the points of intersection of the lines $C'C_1$ and AC, $C'C_2$ and BC, MM' and AC, and MM'' and BC, respectively. By construction, $MO_3 \perp AC$ and $MO_4 \perp BC$, so it follows that right triangles MO_3C and MO_4C have a common hypotenuse MC. This leads to the conclusion that MO_3CO_4 is a cyclic quadrilateral (the sum of its opposite angles is $180°$) whose vertices all lie on a single circle with its diagonal MC as a diameter. Recalling the corollary to the Law of sines that in a triangle the ratio of a side to the sine of the opposite angle is the diameter d of the circumscribed circle, we find that in triangle O_3MO_4,

$$\frac{O_3O_4}{\sin \angle O_3MO_4} = d.$$

Using the fact that MC is the diameter of the circle ($MC = d$), we substitute MC for d:

$$O_3O_4 = MC \sin \angle O_3MO_4. \qquad (1)$$

In a similar way, we can refer to the cyclic quadrilateral $O_1C'O_2C$ and consider triangle $O_1C'O_2$, in which

$$O_1O_2 = CC' \sin \angle O_1C'O_2. \qquad (2)$$

By construction, $MO_3 \parallel C'O_1$ and $MO_4 \parallel C'O_2$, which implies that $\angle O_3MO_4 = \angle O_1C'O_2$.

Since the hypotenuse is the longest side of the right triangle, then in triangle $CC'M$, $CC' < MC$. Referring to (1) and (2), we obtain that $O_1O_2 < O_3O_4$.

Now, observe that O_1, O_2, O_3, and O_4 are the midpoints of $C'C_1$, $C'C_2$, MM', and MM'', respectively. By the Midline Theorem, $O_1O_2 = \frac{1}{2}C_1C_2$ and $O_3O_4 = \frac{1}{2}M'M''$. Returning to the fact that $O_1O_2 < O_3O_4$, we'll get $C_1C_2 < M'M''$.

Recalling that the perimeter of the orthic triangle equals C_1C_2 and that $M'M''$ is less than or equal to the perimeter of the triangle MNK, we see that the desired result is achieved, that is, $C_1C_2 < M'M''$, which means that the triangle of minimal perimeter inscribed in an acute triangle ABC is indeed its orthic triangle. This completes our proof.

Even when the complicated problem is solved, there are usually some questions that remain:

Why did you decide to go in this direction?

How did you come up with that specific idea (in our case, why did we decide to use reflection through an axis, rather than some other transformation of plain)?

How did you come up with the introduction of an auxiliary element? How did you manage to see its usefulness?

How did you manage to separate the essential elements in your diagram to get to the very specific "basic" elements or "basic" figures?

There might be many other questions depending on the problem solved.

So, in conclusion, let's turn back to our steps in the solution process and summarize the logical reasoning behind our ideas.

1. *Identifying the basic elements and their most important attributes*:
   Working with the altitudes in a triangle assumes dealing with several perpendicular segments. This should be a hint to considering the reflection through an axis and utilizing its properties (at least one can try) as one of the possible starting points in contemplating the plan for the proof.
2. *Auxiliary elements*:
   Applying the reflection through an axis allowed us to envision and introduce two auxiliary elements (we will talk about auxiliary elements in detail in one of the subsequent chapters), the segment and the broken line, whose lengths we intended to compare instead of comparing the perimeters of two triangles, the orthic triangle and some randomly inscribed triangle. This shift in strategy made the problem manageable.
3. *Relying on problems-siblings*:
   Recalling the properties of orthic triangle and the properties of a reflection through an axis enabled us to prove that indeed the length of the "substitute" segment is equal to the perimeter of the orthic triangle and the length of the broken line is greater than or equal to the perimeter of a randomly selected triangle inscribed in $\triangle ABC$.
4. *Studying the diagram*:
   Second introduction of auxiliary elements, two cyclic quadrilaterals, was instrumental in finalizing the problem's solution. Even though we did not draw the circle circumscribed about the quadrilaterals MO_3CO_4 and $O_1C'O_2C$ to keep the diagram easier readable, we concentrated on the opposite right angles in these two quadrilaterals (concluding that their sum equals to $180°$), which enlightened each of them as cyclic. This enabled us to identify two inscribed right triangles and refer to the Law of sines to compare their respective legs of interest.
5. *Identifying basic triangles*:
   One more time referring to "basic" triangles, we utilized the property of a midline in a triangle to get the final conclusion comparing C_1C_2 and $M'M''$.

As you study the properties of the orthocenter and the orthic triangle, you discover numerous remarkable facts about the links and connections of various elements in a triangle. They lead to important and interesting generalizations.

It was proved in Theorem 2 that the altitudes of a triangle are the bisectors of the angles of its orthic triangle (Giovanni Fagnano, whose name most often is associated with the above classic problem, deserves another note for being the first to prove that fact as well). If we recall that the center of the inscribed circle in any triangle is the point of intersection of its bisectors, then the orthocenter is the center of the circle inscribed in its orthic triangle. Also, note that the vertices of a triangle are the centers of the excircles (an excircle is a circle lying outside the triangle, tangent to one of its sides and tangent to the extensions of the other two) of its orthic triangle. This becomes obvious after observing that the center of an excircle is the point of intersection of the internal bisector of one angle and the external bisectors of the other two. A set of four points on a plane, where one is the orthocenter of the triangle formed by the other three is called an orthocentric system. I hope that readers looked at Problem 3 at the end of the previous chapter and solved it. We recall it here: *four possible triangles from an orthocentric system all have circumcircles with equal radii.* Another way to prove that is to refer to the nine-point circle of a triangle, which was examined in the previous chapter. Four triangles from orthocentric system all have the same nine-point circle. Readers may wish to explore these properties further and prove that the nine-point circle is tangent to 16 incircles and excircles of the four triangles whose vertices form the orthocentric system.

In conclusion, here are two problems for readers' own solutions (both problems were offered on entrance exams for Moscow State University some years ago):

Problem 7. Let AP and CQ be two altitudes of the acute triangle ABC. Determine the length of the side AC, assuming that the perimeter of the triangle ABC is 15, the perimeter of the triangle BPQ is 9, and the radius of the circumcircle of BPQ is $\frac{9}{5}$.

Problem 8. Let AP and CQ be two altitudes in the acute triangle ABC. The area of ABC is 18, the area of triangle BPQ is 2, and $PQ = 2\sqrt{2}$. Find the radius of the circumcircle of the triangle ABC.

Chapter 4

The Angle Bisector of a Triangle and Its Properties

When you deal with an angle bisector,
don't you recall the protractor?
You get an angle split in half,
as if you slice it with a sharp knife.

This funny small poem gives a vivid interpretation of the definition of an angle bisector:

An angle bisector is the ray passing through a vertex so that it divides the angle into two equal parts.

An important property follows directly from the definition:

A point lies on an angle bisector if and only if it is equidistant from its sides.

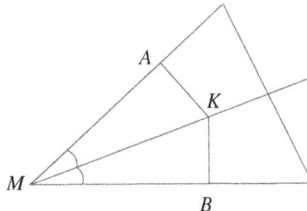

Indeed, if K is any point selected on the angle bisector of the angle with the vertex M, and $KA \perp MA$, $KB \perp MB$, then the congruence of the right triangles MKA and MKB (MK is the common hypotenuse, the adjacent angles are congruent) gives the congruence of the respective legs, $KA = KB$, which is what we were to prove.

To justify the converse statement, we may use the same figure above. Now, given the congruence of perpendiculars dropped from any point K selected on the ray inside the given angle ($KA = KB$), we must prove that MK is the angle bisector of angle M. Once again, we use the congruence of right triangles MKA and MKB (MK is the common hypotenuse, $KA = KB$ as given, and $MA = MB$ by the Pythagorean Theorem). Then the respective angles must be equal, $\angle AMK = \angle BMK$. Therefore, MK is the angle bisector of the angle M.

To restate, an angle bisector is the locus of points equidistant from the sides of the angle, or the locus of points equidistant from the sides of an angle is the angle bisector.

In previous chapters, we started the exploration of special segments of a triangle and showed that medians and altitudes are concurrent. A similar property holds for the three angle bisectors of a triangle as well:

In a triangle, the three angle bisectors intersect at a point — the center of the inscribed circle.

Proof.

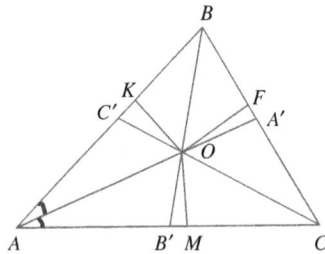

In triangle ABC, AA', BB', and CC' are the angle bisectors.

Denote the point of intersection of AA' and CC' by O. We have to prove that the third bisector BB' passes through O as well.

Draw perpendiculars OK, OM, and OF from O to AB, AC, and BC, respectively. The right triangles OKA and OMA are congruent (by common hypotenuse and adjacent angles). Thus, respective sides are congruent as well, that is, $OK = OM$. Likewise, we get that $OF = OK = OM$. Therefore, O is equidistant from all the sides of the triangle, which implies the concurrency of the three angle bisectors. The point of their intersection is the center of the inscribed circle of the triangle (it is equidistant from all three sides).

As it was the case when we inspected the properties of altitudes and medians, the concurrency of the angle bisectors is a key to the efficient solution of many construction problems. For example, let's look at the following cute, non-standard challenge:

Problem 1. Find the center of the inscribed circle in a triangle with three inaccessible vertices.

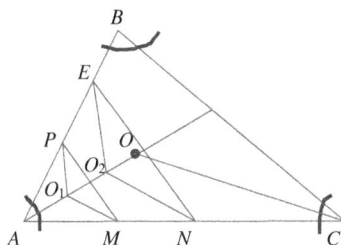

Solution. The goal is to locate the incircle's center. As was just proved, the center of the inscribed circle in the triangle ABC is the point of concurrency of the angle bisectors. To find the center of such a circle, it suffices to construct the angle bisectors of any two angles in the triangle and identify the point of their intersection. Since the vertices A, B, and C cannot be used, we need to locate at least two other points on each angle bisector and then draw the straight lines containing our angle bisectors. The point of their intersection is the solution to the problem.

Even though the vertices of the triangle ABC are not to be used because they are inaccessible, any other points on the sides of the triangle are available. Pick any points P and M on the sides AB and AC, respectively. In the triangle APM, although we can't access vertex A, we can work with the other two vertices. And we can draw the angle bisectors of angles P and M in the triangle APM. Denote the point of their intersection by O_1. The third angle bisector of angle A must pass through O_1 as well. In that way, we found one point on our angle bisector of the angle A. We can repeat the same steps with another pair of arbitrary points E and N selected on AB and AC, respectively. By constructing the two angle bisectors of angles E and N in the triangle EAN, we get the point O_2 where they meet. O_2 must lie on the angle bisector of angle A as well. Therefore, the line passing through O_1 and O_2 contains the angle bisector of angle A. In the same fashion, we can draw the bisector of the angle C (we omit those additional constructions so the picture is easier to understand). The point O, as the point of intersection of the two angle bisectors, is the requested center of incircle of the triangle ABC.

Simply stated, our construction was done and justified by merely applying the fact about the existence of the incenter of a triangle as the point of concurrency of its three angle bisectors.

Now, we will review a few other important properties of the angle bisectors of a triangle.

Theorem 1. *Each angle bisector of a triangle divides the opposite side into segments proportional in length to the adjacent sides.*

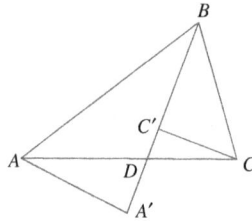

Proof.   In triangle ABC, BD is the bisector of angle B. We have to prove that $AD{:}AB = DC{:}BC$. Let's draw $AA' \perp BD$ and $CC' \perp BD$. Triangles $AA'D$ and $CC'D$ are similar, because $\angle AA'D = \angle CC'D = 90°$ and $\angle ADA' = \angle CDC'$ as vertical angles. Therefore,

$$AD{:}AA' = DC{:}CC'. \qquad (1)$$

Triangles $AA'B$ and $CC'B$ are also similar because $\angle ABA' = \angle CBC'$ and $\angle AA'B = \angle CC'B = 90°$. Hence,

$$AB{:}AA' = BC{:}CC'. \qquad (2)$$

By dividing (1) by (2), we obtain the desired result, $AD{:}AB = DC{:}BC$.

The next theorem provides a helpful relationship that can be used to find the length of an angle bisector of a triangle when you know two sides and the angle that they form. The formula will be used in a few problems later on.

Theorem 2. *In triangle ABC, given $AC = b$, $AB = c$, and AD is the bisector of angle A. Prove that $AD = \frac{2bc}{b+c} \cdot \cos(A/2)$.*

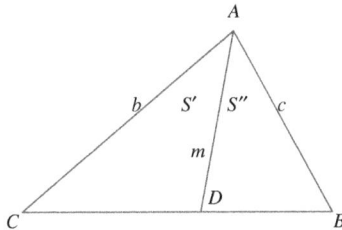

Proof. Let's denote for convenience the area of the triangle ADC by S', the area of triangle ADB by S'', and the length of the angle bisector $AD = m$. Then

$$S' = \frac{1}{2}bm \cdot \sin(A/2), \text{ and } S'' = \frac{1}{2}cm \cdot \sin(A/2).$$

The area of the triangle ABC, $S = \frac{1}{2} bc \cdot \sin A$.
On the other hand, $S = S' + S''$. It follows that

$$bc \cdot \sin A = bm \cdot \sin\left(\frac{A}{2}\right) + cm \cdot \sin\left(\frac{A}{2}\right).$$

Using the formula for sine of a double angle, $\sin 2\alpha = 2\sin \alpha \cos \alpha$, gives $2bc \cdot \sin(A/2)\cos(A/2) = (b+c)m \cdot \sin(A/2)$; and after dividing both sides by $\sin(A/2) \neq 0$ we arrive at $2bc \cdot \cos(A/2) = (b + c)m$.
From the last equality, we derive that $m = \frac{2bc}{b+c} \cdot \cos(A/2)$, which was to be proved.

Theorem 3. *In triangle ABC, AM and BD are the bisectors of the angles A and B, respectively, and O is the point of their intersection. Prove that $BO{:}OD = (AB + BC){:}AC$.*

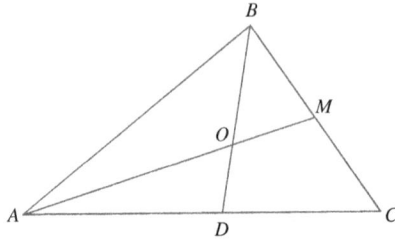

Proof. By Theorem 1, in the triangle ABC, $DC{:}DA = BC{:}BA$. This can be written as $1 + (DC/DA) = 1 + (BC/BA)$, or $(AD + DC)/AD = (AB + BC)/AB$. Noticing that $AD + DC = AC$, we get

$$AC/AD = (AB + BC)/AB, \text{ or } AD = AC \cdot AB/(AB + BC). \quad (1)$$

By Theorem 1, in the triangle BAD,

$$BO{:}OD = AB{:}AD. \quad (2)$$

Substituting (1) into (2), we obtain that $BO{:}OD = (AB + BC){:}AC$, which is the desired equality.

Theorem 4. *In triangle ABC, AD is the bisector of the angle A, $AC = b, AB = c, AD = m, CD = b'$, and $BD = c'$.*
    Prove that $m^2 = bc - b'c'$.

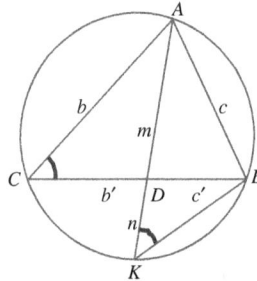

Proof. To simplify the proof, we will draw the circumscribed circle of the triangle ABC. Assume the straight line AD meets the circle at the point K. It's easy to see that triangles AKB and ACD are similar. Indeed, $\angle AKB = \angle ACD$ (angles AKB and ACB both are subtended by arc AB), and $\angle CAD = \angle KAB$, since AD is the bisector of the angle A. Then $AK{:}AC = AB{:}AD$. If we put $DK = n$, then

$$(n + m){:}b = c{:}m, \text{ from which } m^2 = bc - mn. \quad (*)$$

From the Intersecting chords theorem (the products of the lengths of the line segments on each chord are equal), we know that $nm = b'c'$. Substituting this into (*) yields the required formula $m^2 = bc - b'c'$.

There are several important and useful relationships that emanate from the above theorems.

Problem 2. In a right triangle, the bisector of an acute angle divides the opposite side into segments of the length m and n. Find the length of each side of the triangle.

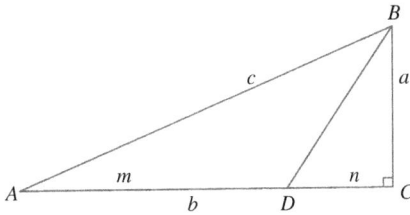

Solution. In triangle ACB, $\angle C = 90°$, BD is the bisector of the angle B, $AD = m$, and $DC = n$. Let $AB = c$, $AC = b$, and $BC = a$.

First, we notice that $AC = m + n$. Next, by Theorem 1, in triangle ABC, $AB{:}BC = AD{:}DC$, or $AB{:}BC = m{:}n$, from which

$$c = \frac{m}{n} \cdot a. \tag{1}$$

Substituting (1) into $c^2 = a^2 + b^2$, which holds due to the Pythagorean Theorem, we obtain $(m+n)^2 + a^2 = \left(\frac{m}{n}\right)^2 \cdot a^2$ or equivalently, $a^2 \left(\left(\frac{m}{n}\right)^2 - 1\right) = (m+n)^2$.

After simplification, we see that $a = n\sqrt{\frac{m+n}{m-n}}$. Finally, recalling the property of the bisector that $c = \frac{m}{n} \cdot a$, $(m > 0, n > 0)$ and substituting the value of a, we arrive at

$$c = \sqrt{\left(\frac{m}{n}\right)^2 \cdot \frac{m+n}{m-n} \cdot n^2} = m\sqrt{\frac{m+n}{m-n}}.$$

Problem 3. Prove that if in a triangle ABC, $\angle A = 120°$, k is the bisector of the given angle A, and b and c are the sides that form that angle, then $\frac{1}{k} = \frac{1}{b} + \frac{1}{c}$.

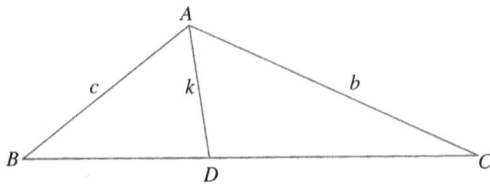

Proof. By Theorem 2,

$k = \frac{2bc}{b+c} \cdot \cos\left(\frac{A}{2}\right) = \frac{2bc}{b+c} \cdot \cos 60° = \frac{2bc}{b+c} \cdot \frac{1}{2} = \frac{bc}{b+c}$. From this equality, we arrive at $\frac{1}{k} = \frac{1}{b} + \frac{1}{c}$.

Isn't it amazing how easily this relationship evolved from just one formula, the formula proved in Theorem 2?!

It is easy to prove that in an isosceles triangle, two of its angle bisectors by its base are congruent. It follows immediately from the congruence of the two triangles formed by the common base of an isosceles triangle and the two bisectors of the angles adjacent to that base. We leave the justification to the reader.

The proof of the opposite statement, however, is not that trivial and is well-known as *Steiner–Lehmus Theorem*, which will be examined below as our Problem 4.

Problem 4 (Steiner–Lehmus Theorem). Prove that any triangle with two congruent angle bisectors is isosceles.

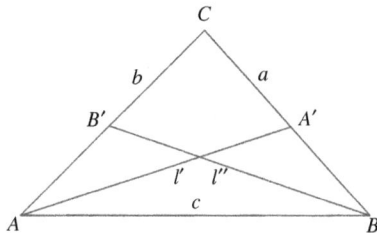

Proof. Assume $l' = l''$, where l' and l'' are given equal angle bisectors of the triangle ABC. We have to prove that $BC = AC$ or, which is the same, that $\angle A = \angle B$.

Let's denote $BC = a$, $AC = b$, and $AB = c$. By Theorem 2,

$$l' = \frac{2bc}{b+c} \cdot \cos\left(\frac{A}{2}\right) \text{ and } l'' = \frac{2ac}{a+c} \cdot \cos\left(\frac{B}{2}\right).$$

If $l' = l''$, then $\frac{bc}{b+c} \cdot \cos\left(\frac{A}{2}\right) = \frac{ac}{a+c} \cdot \cos\left(\frac{B}{2}\right).$ (*)

Suppose $a > b$. Then for (*) to be true it's obvious that $\cos\left(\frac{A}{2}\right)$ has to be greater than $\cos\left(\frac{B}{2}\right)$, or angle A has to be less than angle B, which contradicts our assumption that $a > b$ (if angle A is less than angle B, then the corresponding opposite side a has to less than b). In the same manner, a cannot be less than b. We have arrived at a contradiction, and this proves our assertion. Therefore, $a = b$.

Problem 5. The sides of the triangle ABC equal a, b, and c. Find the length of each angle bisector of the triangle.

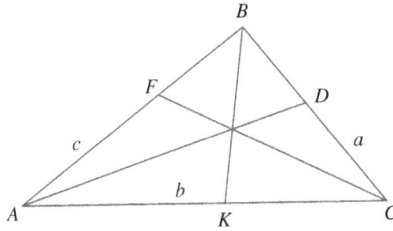

Solution. In triangle ABC, $AB = c$, $BC = a$, and $AC = b$. BK, CF, and AD are the bisectors of the angles B, C, and A, respectively. By Theorem 1, $\frac{AK}{KC} = \frac{c}{a}$. Thus, $AK = KC \cdot \frac{c}{a}$. Remembering that $AK + KC = b$, we easily find that $KC = \frac{ab}{a+c}$.

Hence $AK = \frac{cb}{a+c}$.

By Theorem 4,

$$BK = \sqrt{AB \cdot BC - AK \cdot KC} = \sqrt{ac - \frac{ab}{a+c} \cdot \frac{bc}{a+c}}$$

$$= \sqrt{ac - \frac{acb^2}{(a+c)^2}} = \sqrt{ac\left(1 - \left(\frac{b}{a+c}\right)^2\right)} = \sqrt{\frac{ac((a+c)^2 - b^2)}{(a+c)^2}}$$

$$= \frac{\sqrt{ac(a+c-b) \cdot (a+c+b)}}{a+c}$$

(we omitted the absolute value sign in denominator because, clearly, under the given conditions, $a + c > 0$).

In the same way, it's possible to find CF and AD. These calculations are left to the readers.

Problem 6. In triangle ABC, BK is the bisector of the angle B and $BK = BC$. O is the point of intersection of the angle bisectors of the triangle, and $BO{:}OK = 2 : 1$. The perimeter of the triangle ABC is 21. Find its sides.

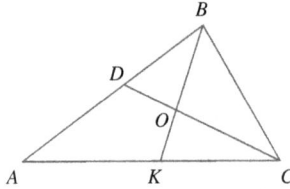

Solution. By Theorem 3, $BO{:}OK = (AB + BC){:}AC$. We know that $BO{:}OK = 2 : 1$ and the perimeter equals 21 (i.e., $AB + BC + AC = 21$). Therefore, $\frac{21 - AC}{AC} = 2$. Solving this simple equation, we get the first side, $AC = 7$.

Now, let's consider triangle KCB. CO is the bisector of the angle C and therefore, by Theorem 1,

$$BO{:}OK = BC{:}CK. \text{ So, } BC = 2CK. \tag{1}$$

In triangle ACB,

$$AK{:}AB = CK{:}BC, \text{ from which } AB = 2AK. \tag{2}$$

By Theorem 4,

$$BK^2 = AB \cdot BC - AK \cdot CK. \tag{3}$$

Recalling that by the conditions of the problem, $BK = BC$ and substituting (1) and (2) into (3) we obtain

$$BC^2 = AB \cdot BC - \frac{1}{2}AB \cdot \frac{1}{2}BC,$$

$$BC^2 = \frac{3}{4}AB \cdot BC, \text{ from which } BC = \frac{3}{4}AB.$$

We have already found side AC, $AC = 7$. The perimeter of the triangle is 21, hence $AB + BC = 14$. So, $AB + \frac{3}{4}AB = 14$, leading to $AB = 8$. Finally, we find that $BC = \frac{3}{4} \cdot 8 = 6$.

As we compared a few properties of a triangle's medians, altitudes, and angle bisectors, it has to be noted that the angle bisectors are perhaps the most troublesome. Even the problem of constructing a triangle given its three angle bisectors using only a compass and straightedge is impossible to solve, while similar constructions of a triangle given its three medians or three altitudes are solvable — interesting construction problems to investigate on your own.

There are many more challenging problems involving properties of angle bisectors. We have concentrated on a few that exhibit the formulas and relationships between sides, angles (their trigonometric functions), and internal angle bisectors in a triangle. We invite the reader to further explore this interesting topic and extend it to the examination of external angle bisectors (lines bisecting the angles formed by the sides of a triangle and their extensions).

We will conclude this chapter by offering a few problems for you to practice.

Problem 7. Prove that if in a triangle one angle equals $120°$, then the triangle with vertices in the feet of the bisectors of the given triangle is a right triangle (see figure below).

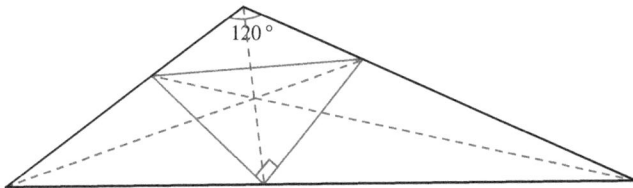

We encourage the readers to solve Problem 7 *before* reading the next chapters. It will be interesting to compare your solution to the one provided in the book. You perhaps will have then more appreciation for the methods introduced later on.

Problem 8. The legs of a right triangle equal 3 and 4, respectively. Find the bisector of the right angle.

Problem 9. In triangle ABC, $AB = BC = a$, $AC = b$, and AN and CM are the bisectors of the angles A and C, respectively. Find the length of the segment MN.

Chapter 5

The Area of a Quadrilateral

Historically, many of the most important geometrical concepts arose from problems that are practical in origin. A good example is the calculation of the area of convex polygons. The area of the triangle as a basic plane polygon is often employed in calculating the area of other polygons. It is most commonly calculated by the formulas $S = \frac{1}{2}ah$ (half the product of the base by the altitude dropped to that base) and $S = \frac{1}{2}ab \cdot \sin\gamma$ (half the product of two sides and the sine of the angle between them), which we have already used a few times.

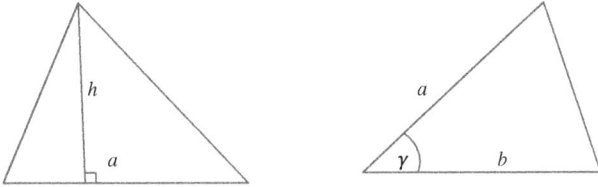

The area of a triangle in terms of the sides' lengths is found by *Heron's formula*, which was first mentioned and proved in the book *Metrica* by the prominent Greek mathematician, engineer, and inventor Heron of Alexandria (c. 10–c. 70 AD):

$$S = \sqrt{p(p-a)(p-b)(p-c)},$$

where a, b, and c are the sides, and p is the semi-perimeter of a triangle.

While the first two formulas are covered in high school, we can't say the same about Heron's formula. Even though the main focus of this chapter is on the areas of quadrilaterals, Heron's formula deserves mentioning here. First of all, you can't ignore and not emphasize its resemblance to *Brahmagupta's formula*, which will be studied in this chapter. Secondly, the introduction of this classic formula and its proof will help motivate the development of the formulas for the area of quadrilaterals and the ideas behind the proofs to follow.

Heron's original proof made use of the properties of cyclic quadrilaterals, one of which we will consider below (we'll also use it as a very powerful tool in some problems examined in the later chapters):

A quadrilateral can be inscribed in a circle if and only if a pair of opposite angles is supplementary.

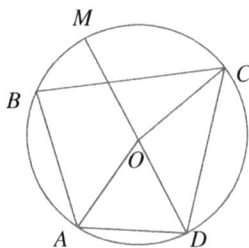

For proof of the direct statement, we will recall that the measure of an inscribed angle in a circle is half the measure of the corresponding central angle:

$$\angle ABC = \frac{1}{2}\angle AOC, \text{ or } 2\angle ABC = \angle AOC. \qquad (*)$$

Draw the diameter DM. $AO = DO = OM$ as radii of the same circle. Therefore, the triangle AOD is isosceles, and $2\angle ADO = 180° - \angle AOD$, since $\angle ADO = \angle DAO$. Analogously, in the isosceles triangle DOC, $2\angle CDO = 180° - \angle DOC$, since $\angle CDO = \angle DCO$.

By adding the equalities, we obtain
$2\angle ADO + 2\angle CDO = 360° - \angle AOD - \angle DOC$, or
$2(\angle ADO + \angle CDO) = 360° - (\angle AOD + \angle DOC)$, which leads to
$2\angle ADC = 360° - \angle AOC$. After substituting angle AOC from $(*)$

and dividing both sides by 2, we get the final result:

$$\angle ADC = 180° - \angle ABC, \text{ or } \angle ADC + \angle ABC = 180°,$$

and expressing this in words,

If a quadrilateral is inscribed in a circle, its opposite angles are supplementary.

Let's now consider a quadrilateral $ABCD$ whose opposite angles are supplementary: $\angle A + \angle C = \angle D + \angle B$. We have to prove that $ABCD$ is a cyclic quadrilateral.

Draw a circumcircle about the triangle ABC and assume that it intersects AD at some point D' different from D. The quadrilateral $ABCD'$ is cyclic and the pairs of its opposite angles are supplementary, $\angle A + \angle C = \angle D' + \angle B$.

By transitivity, $\angle B + \angle D' = \angle B + \angle D$, which implies $\angle D' = \angle D$. We got the contradiction to the assumption that points D and D' do not coincide, which proves the fact that

If the opposite angles of a quadrilateral are supplementary, then it is a cyclic quadrilateral.

The above property allows us to arrive at an interesting proof of Heron's formula.

In triangle ABC, let $AB = c$, $BC = a$, $AC = b$, and $p = \frac{1}{2}(a + b + c)$. Denote by O the center of the incircle of triangle ABC. The perpendiculars dropped from O to the sides of ABC are equal as radii of the same circle: $ON = OL = OK$.

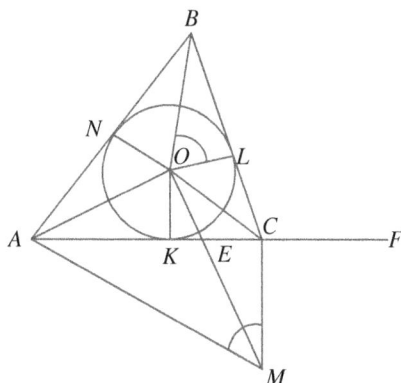

Recalling that O is the point of intersection of the angle bisectors of the triangle ABC, it's easy to prove the congruence of the pairs of right triangles OKA and ONA, ONB and OLB, and OLC and OKC. The pairs of corresponding sides in those triangles are equal: $AK = AN$, $NB = BL$, and $LC = CK$. If we now locate point F on the extension of AC such that $CF = BL$, then the length of AF is equal to the semi-perimeter of the triangle ABC. Indeed,

$$p = \frac{1}{2}(AB + BC + CA) = \frac{1}{2}(AN + NB + BL + LC + CK + KA)$$

$$= \frac{1}{2}(2AK + 2KC + 2BL) = AK + KC + CF = AF.$$

In Chapter 2, it was proved that the area of a triangle is $S = \frac{1}{2}Pr$, where P is the triangle's perimeter and r is its inradius. Therefore, the area of triangle ABC equals

$$S = pr = OK \cdot AF. \tag{1}$$

The next step is to construct a perpendicular to AO at O and a perpendicular to AF at C. Denote the point of their intersection by M. The right triangles AOM and ACM share the common hypotenuse AM. Hence, both triangles are inscribed in a common circle with AM as its diameter. It follows then that the quadrilateral $AOCM$ is cyclic, implying that angles AOC and AMC are supplementary. Noting that the sum of pairs of equal angles at point O is $360°$, we have

$$\angle AOB + \angle BOC + \angle COA = 2\angle AOK + 2\angle BOL + 2\angle COK = 360°$$

and respectively,

$$\angle AOK + \angle BOL + \angle COK = 180°.$$

But $\angle AOK + \angle COK = \angle AOC$, therefore, $\angle AOC + \angle BOL = 180°$, which leads to a conclusion that $\angle BOL = \angle AMC$ as supplementary to the same angle AOC (recall that angles AOC and AMC are supplementary in the cyclic quadrilateral $AOCM$). In the right triangles ACM and BLO, two respective angles are equal, therefore they are similar. If we denote by E the point of intersection of OM and AC, then the right triangles OKE and MCE are similar as well ($\angle OKE = \angle MCE = 90°$ and $\angle OEK = \angle MEC$ as vertical angles).

From the pairs of similar triangles we obtain that the ratios of the respective sides are equal: $AC/BL = MC/OL$ and $MC/OK = CE/EK$. Recalling that $OL = OK = r$ and dividing one equality by the other, we get that $AC/BL = CE/EK$. Substituting BL for CF (the segments are of the equal length by construction), we conclude that $AC/CF = CE/EK$.

Adding 1 to both sides and doing a few simple algebraic manipulations yield $AC/CF + 1 = CE/EK + 1$, so $(AC + CF)/CF = (CE + EK)/EK$, which implies that $AF/CF = CK/EK$. By multiplying the numerator and denominator by the same number AF on the right side and by AK on the left side, we get

$$AF^2/(CF \cdot AF) = (CK \cdot AK)/(EK \cdot AK). \qquad (2)$$

Let's now consider the right triangle AOE (AO and OE are perpendicular by construction). Recall that the length of the altitude on the hypotenuse of a right triangle is the geometrical mean between the lengths of the segments of the hypotenuse:

$$AK \cdot EK = OK^2. \qquad (3)$$

Substituting (3) into (2), we get that $AF^2/(CF \cdot AF) = (CK \cdot AK)/OK^2$, from which $AF^2 \cdot OK^2 = CF \cdot AF \cdot CK \cdot AK$ leading to $AF \cdot OK = \sqrt{CF \cdot AF \cdot CK \cdot AK}$.

From (1) we conclude that the area of triangle ABC is

$$S = \sqrt{CF \cdot AF \cdot CK \cdot AK}.$$

We showed before that $AF = p$. Obviously, $CF = AF - AC = p - b$, $CK = AF - (AK + CF) = p - c$, and $AK = AF - (KC + CF) = p - a$.

Therefore, substituting respective values, the area of triangle ABC is modified to $S = \sqrt{p(p-a)(p-b)(p-c)}$, and we are done.

At this point, we are ready to move on from triangles to quadrilaterals.

While the calculation of the area of a triangle is presented in high school in a number of ways, as we mentioned earlier, for convex quadrilaterals there are area formulas only for parallelograms and trapezoids.

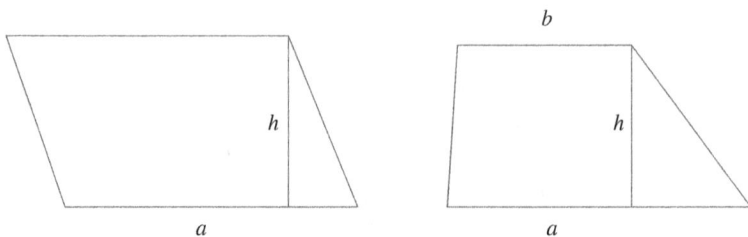

Area of a parallelogram $S = ah$ (a is the base, and h is the altitude dropped to it).

Area of a trapezoid $S = \frac{1}{2}(a+b)h$ (a and b are the bases and h is the altitude).

What about the general formula for the area of any convex quadrilateral? Wouldn't it be great to have it in your arsenal as a useful tool to solve area problems? Such a formula first was discovered by German mathematician Carl Anton Bretschneider (1808–1878) in 1842. It's interesting to note that another German mathematician, Karl Georg Christian von Staudt (1798–1867) got the same result independently in the same year.

In this chapter, we will introduce Bretschneider's formula along with a few corollaries and demonstrate their practical application in problem solving. The techniques (proposed formulas) are interesting and timesaving, and can also be useful in giving a feel for the estimation of answers.

We will apply some trigonometry to help us develop a formula in the following main theorem. We assume readers' familiarity with trigonometric formulas that will be used in the proof, such as $\cos^2 x + \sin^2 x = 1$, $\cos(x + y) = \cos x \cos y - \sin x \sin y$, and $1 + \cos x = 2\cos^2\left(\frac{x}{2}\right)$.

Theorem 1. *The area of a convex quadrilateral is*

$$S = \sqrt{(p-a)(p-b)(p-c)(p-d) - abcd \cdot \cos^2\left(\frac{\beta + \eth}{2}\right)},$$

where a, b, c, d are the sides of the quadrilateral $ABCD$, $\beta = \angle ABC$, $\eth = \angle ADC$, and $p = \frac{a+b+c+d}{2}$.

In all the following theorems and problems, we will denote by a, b, c, and d the sides and by p the semi-perimeter of a quadrilateral, and by β and $\eth$ its opposite angles (unless it is specifically indicated otherwise).

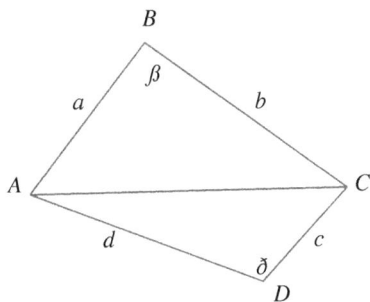

Proof. Consider two triangles ABC and ADC. By the Law of cosines,

$$AC^2 = a^2 + b^2 - 2ab\cos\beta \text{ (from triangle } ABC\text{)};$$
$$AC^2 = c^2 + d^2 - 2cd\cos\eth \text{ (from triangle } ADC\text{)}.$$

Then $a^2 + b^2 - 2ab\cos\beta = c^2 + d^2 - 2cd\cos\eth$, or equivalently,

$$a^2 + b^2 - c^2 - d^2 = 2ab\cos\beta - 2cd\cos\eth. \tag{1}$$

The area of $ABCD$ is equal to the sum of the areas of triangles ABC and ADC: $S = \frac{1}{2}ab\sin\beta + \frac{1}{2}cd\sin\eth$. Hence,

$$4S = 2ab\sin\beta + 2cd\sin\eth. \tag{2}$$

Square (1) and (2) and add:

$$(a^2 + b^2 - c^2 - d^2)^2 + 16S^2 = (2ab\cos\beta - 2cd\cos\eth)^2$$
$$+ (2ab\sin\beta + 2cd\sin\eth)^2,$$
$$(a^2 + b^2 - c^2 - d^2)^2 + 16S^2 = 4a^2b^2(\cos^2\beta + \sin^2\beta)$$
$$+ 4c^2d^2(\cos^2\eth + \sin^2\eth)$$
$$- 8abcd(\cos\beta\cos\eth - \sin\beta\sin\eth),$$
$$(a^2 + b^2 - c^2 - d^2)^2 + 16S^2 = 4a^2b^2 + 4c^2d^2 - 8abcd\cos(\beta + \eth),$$
$$16S^2 = (4a^2b^2 + 4c^2d^2 + 8abcd) - 8abcd - 8abcd\cos(\beta + \eth) - (a^2 + b^2 - c^2 - d^2)^2.$$

Modifying the right-hand side, we get

$$(2ab + 2cd)^2 - 8abcd - 8abcd\cos(\beta + \delta) - (a^2 + b^2 - c^2 - d^2)^2$$
$$= (2ab + 2cd)^2 - (a^2 + b^2 - c^2 - d^2)^2 - 8abcd(1 + \cos(\beta + \delta))$$
$$= (2ab + 2cd - a^2 - b^2 + c^2 + d^2)(2ab + 2cd + a^2 + b^2 - c^2 - d^2)$$
$$- 16abcd \cdot \cos^2\left(\frac{\beta + \delta}{2}\right)$$
$$= ((c + d)^2 - (a - b)^2)((a + b)^2 - (c - d)^2) - 16abcd \cdot \cos^2\left(\frac{\beta + \delta}{2}\right)$$
$$= (c + d + b - a)(c + d + a - b)(a + b + d - c)(a + b + c - d)$$
$$- 16abcd \cdot \cos^2\left(\frac{\beta + \delta}{2}\right).$$

After simplifying and noting that $2(p - a) = 2p - 2a = 2(a + b + c + d)/2 - 2a = a + b + c + d - 2a = b + c + d - a$ (the analogous equality holds for each side of the quadrilateral), we can rewrite the right-hand side of the last equality as

$$2(p - a)2(p - b)2(p - c)2(p - d) - 16abcd \cdot \cos^2\left(\frac{\beta + \delta}{2}\right).$$

Therefore,

$$16S^2 = 2(p - a)2(p - b)2(p - c)2(p - d) - 16abcd \cdot \cos^2\left(\frac{\beta + \delta}{2}\right).$$

Dividing both sides by 16, we finally arrive at

$$S = \sqrt{(p - a)(p - b)(p - c)(p - d) - abcd \cdot \cos^2\left(\frac{\beta + \delta}{2}\right)},$$

which is what we wished to prove.

This theorem has three important corollaries.

The first one bears the name of the Indian mathematician Brahmagupta (598 CE–670 CE), whose contributions to the field of mathematics were quite substantial. He was the first to introduce arithmetic operations with 0, give general solutions to the linear and quadratic equations, and perform operations with fractions and negative numbers, to name just a few. In geometry, his name is associated with the formula for the calculation of the area of cyclic quadrilaterals. It is amazing that in his works, he did not provide any proof or even a hint as to how the formula was derived, so it is not clear how he got the result. In our exploration of his formula, the proof will be based on the results of the general formula from Theorem 1.

Theorem 2 (Brahmagupta's formula). *If a quadrilateral is cyclic, then its area is*

$$S = \sqrt{(p-a)(p-b)(p-c)(p-d)}.$$

Proof. As we proved at the beginning of the chapter, in a cyclic quadrilateral opposite angles are supplementary.

So, we see that $\beta + \eth = 180°$ and therefore, $\cos^2\left(\frac{\beta+\eth}{2}\right) = 0$.

By Theorem 1, we obtain the desired result

$$S = \sqrt{(p-a)(p-b)(p-c)(p-d)}.$$

Theorem 3. *If a quadrilateral is circumscribed about a circle, its area is*

$$S = \sqrt{abcd \cdot \sin^2\left(\frac{\beta+\eth}{2}\right)}.$$

Proof. We begin by proving that in any circumscribed quadrilateral (tangential quadrilateral) two sums of the pairs of opposite sides are equal. This is known as *Pitot's Theorem*, named after the French engineer Henri Pitot (1695–1771).

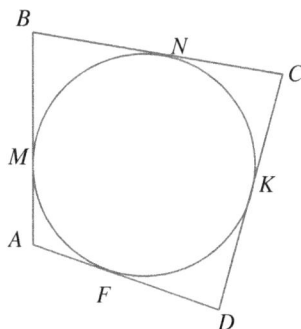

For the proof, it will suffice to recall that two tangent line segments from a point outside a circle have equal lengths: $AM = AF$,

$BM = BN$, $CK = CN$, and $DK = DF$. Adding we obtain $AM + BM + CK + DK = AF + BN + CN + DF$. Regrouping gives $(AM + BM) + (CK + DK) = (AF + DF) + (BN + CN)$, which yields $AB + CD = AD + BC$.

Hence, $a + c = b + d$, which is the relationship we set to develop.

In light of this, recalling that the semi-perimeter $p = \frac{a+b+c+d}{2}$, and using the above equality, it's easy to obtain that $p - a = c$, $p - b = d$, $p - c = a$, and $p - d = b$.

By Theorem 1, we get the desired formula:

$$S = \sqrt{abcd - abcd \cdot \cos^2\left(\frac{\beta + \eth}{2}\right)} = \sqrt{abcd \cdot \left(1 - \cos^2\left(\frac{\beta + \eth}{2}\right)\right)}$$

$$= \sqrt{abcd \cdot \sin^2\left(\frac{\beta + \eth}{2}\right)}.$$

Theorem 4. *If a quadrilateral is inscribed in a circle and is circumscribed about the circle simultaneously, its area is the square root of the product of its sides:*

$$S = \sqrt{abcd}.$$

Proof. Armed with Theorems 1 through 3, we don't even need to display the diagram to solve this problem.

By Theorem 2,

$$S = \sqrt{(p - a)(p - b)(p - c)(p - d)},$$

and we know that $a + c = b + d$ (the sum of the opposite sides of the circumscribed quadrilateral).

Therefore,

$$S = \sqrt{\frac{1}{2}(c + d + b - a) \cdot \frac{1}{2}(c + d + a - b) \cdot \frac{1}{2}(a + b + d - c) \cdot \frac{1}{2}(a + b + c - d)}$$

$$= \sqrt{(2/2)c \cdot (2/2)d \cdot (2/2)a \cdot (2/2)b} = \sqrt{abcd}.$$

It is interesting to note that Heron's formula for the area of a triangle may be derived as a corollary from Brahmagupta's formula. If $d = 0$, the cyclic quadrilateral becomes a triangle and then its area will be

$$S = \sqrt{p(p-a)(p-b)(p-c)}.$$

Usually, Heron's formula is applied when the only information that is given about a triangle is the lengths of its sides. What about the formulas in this chapter? When would you make use of them?

Most likely the formulas would be handy in approaching problems where a quadrilateral has a given set of sides or opposite angles. You might get unexpected elegant and beautiful solutions simply by applying the discussed formulas. Let's look at some examples.

Problem 1. Of all quadrilaterals inscribed in a circle, find the one with the greatest area.

Solution. For the proof, we will use classic AM-GM Inequality, the property that the arithmetic mean of two positive integers is always greater than or equal to their geometric mean: for $x > 0$, $y > 0$, $\frac{x+y}{2} \geq \sqrt{xy}$ with equality possible if and only if $x = y$.

If a quadrilateral is a cyclic one and p is its semi-perimeter, then its area

$$\begin{aligned} S &= \sqrt{(p-a)(p-b)(p-c)(p-d)} \\ &= \sqrt{(p-a)(p-b)} \cdot \sqrt{(p-c)(p-d)} \\ &\leq \tfrac{1}{2}((p-a)+(p-b)) \cdot \tfrac{1}{2}((p-c)+(p-d)) \\ &= \tfrac{1}{4}(c+d)(a+b) \leq \tfrac{1}{16}(a+b+c+d)^2 = \left(\frac{P}{4}\right)^2, \end{aligned}$$

where P is the perimeter.

Note now that $S = (\frac{P}{4})^2$ only when $a = b = c = d$. For a quadrilateral inscribed in a circle, that's possible only when it is a square. Therefore, of all cyclic quadrilaterals, a square will have the greatest area.

It should be fairly easy now for readers to prove another interesting statement, which follows directly from the above problem:

Of all quadrilaterals of a given perimeter, a square has the greatest area.

Problem 2. In a convex quadrilateral $ABCD$, the lengths of the sides are 2, 2, 4, and 6. Its area is $5\sqrt{3}$. Prove that there exists a circle circumscribed about $ABCD$.

68   *Geometrical Kaleidoscope (Second Edition)*

Solution. By the general formula from Theorem 1,

$$S = \sqrt{(p-a)(p-b)(p-c)(p-d) - abcd \cdot \cos^2\left(\frac{\beta + \eth}{2}\right)}.$$

First, we find p as $p = (2+2+4+6)/2 = 7$. Next, after substituting all the values into the formula for the area of a quadrilateral, we'll get that

$$5\sqrt{3} = \sqrt{5 \cdot 5 \cdot 3 \cdot 1 - 2 \cdot 2 \cdot 4 \cdot 6 \cos^2\left(\frac{\beta + \eth}{2}\right)},$$

$$5\sqrt{3} = \sqrt{75 - 96 \cos^2\left(\frac{\beta + \eth}{2}\right)}.$$

Squaring both sides leads to $75 = 75 - 96\cos^2\left(\frac{\beta+\eth}{2}\right)$, from which $96\cos^2\left(\frac{\beta+\eth}{2}\right) = 0$.

It follows that $\cos^2\left(\frac{\beta+\eth}{2}\right) = 0$ resulting in $\cos\left(\frac{\beta+\eth}{2}\right) = 0$.

Therefore, $\frac{\beta+\eth}{2} = 90°$ leading to $\beta + \eth = 180°$. We got that the opposite angles of the quadrilateral are supplementary, which means it is a cyclic quadrilateral and there exists a circle circumscribed about it, and our proof is complete.

Problem 3. Find the area of a convex quadrilateral with sides 2, 3, 3, 4, and opposite pair of angles 63° and 117°.

Solution. The sum of the opposite angles is $63° + 117° = 180°$ making the given quadrilateral a cyclic quadrilateral. Its semi-perimeter $p = \frac{1}{2}(2+3+3+4) = 6$, and respectivly, $p - 2 = 4$, $p - 3 = 3$, and $p - 4 = 2$.

Then by the formula from Theorem 2, we have $S = \sqrt{4 \cdot 3 \cdot 3 \cdot 2} = 6\sqrt{2}$.

Problem 4. $ABCD$ is a tangential quadrilateral. The sum of the opposite angles of $ABCD$ is 120°. Its area is $\frac{\sqrt{3}}{2}$. Prove that the product of its sides is 1.

Proof. If $ABCD$ is a quadrilateral circumscribed about a circle, then its area

$$S = \sqrt{abcd \cdot \sin^2\left(\frac{\beta + \eth}{2}\right)} = \sqrt{abcd \cdot \sin^2\left(\frac{120°}{2}\right)} = \sqrt{\frac{3}{4}abcd}.$$

Since $S = \frac{\sqrt{3}}{2}$, $\frac{\sqrt{3}}{2} = \sqrt{\frac{3}{4}abcd}$, and after squaring both sides, we have $\frac{3}{4} = \frac{3}{4}abcd$, from which we arrive at the desired result, $abcd = 1$.

Problem 5. What is the maximum area of a convex quadrilateral with sides 1, 4, 7, and 8?

Solution. There are many different quadrilaterals with the given sides. They will differ from each other by the angles between sides. The area of any such quadrilateral may be calculated by the general formula

$$S = \sqrt{(p - a)(p - b)(p - c)(p - d) - abcd \cdot \cos^2\left(\frac{\beta + \eth}{2}\right)}.$$

Obviously, $S \leq \sqrt{(p - a)(p - b)(p - c)(p - d)}$, with equality when $\cos^2\left(\frac{\beta + \eth}{2}\right) = 0$. That occurs only when $\beta + \eth = 180°$, which means that the maximum area will be found for a cyclic quadrilateral.
  Let's find the value of p: $p = \frac{1}{2}(1 + 4 + 7 + 8) = 10$. Then $p - 1 = 9$, $p - 4 = 6$, $p - 7 = 3$, and $p - 8 = 2$.
  Substituting the numbers into the formula for the area, we get $S = \sqrt{9 \cdot 6 \cdot 3 \cdot 2} = 18$.

Problem 6. The lengths of the sides of a cyclic quadrilateral form an arithmetic sequence with the common difference 2. Its area is $\sqrt{105}$. Find the length of each side.

Solution. Denote the length of the smallest side of the quadrilateral by $a_1 = x$. Then its sides will have the lengths, in ascending order

(by the definition of arithmetic sequence)

$$a_2 = x + 2,$$
$$a_3 = x + 4,$$
$$a_4 = x + 6.$$

The semi-perimeter

$$p = \frac{1}{2}(x + (x + 2) + (x + 4) + (x + 6)) = \frac{1}{2}(4x + 12) = 2x + 6.$$

Then $p - a_1 = x + 6$, $p - a_2 = x + 4$, $p - a_3 = x + 2$, and $p - a_4 = x$. The area of the cyclic quadrilateral is

$$S = \sqrt{(p - a_1)(p - a_2)(p - a_3)(p - a_4)}.$$

After substituting the value of S and the values of the factors, we get the equation

$$\sqrt{105} = \sqrt{(x + 6)(x + 4)(x + 2)x},$$
$$(x + 6)(x + 4)(x + 2)x = 105,$$
$$((x + 6)x)((x + 4)(x + 2)) = 105,$$
$$(x^2 + 6x)(x^2 + 6x + 8) = 105.$$

To solve this equation, we will introduce another variable:

$$y = x^2 + 6x, \tag{1}$$

which will transform the original equation into $y(y + 8) = 105$, or equivalently, $y^2 + 8y - 105 = 0$.

Let's solve this quadratic equation.

$$D = 8^2 + 4 \cdot 105 = 484,$$

$$y = \frac{-8 \pm \sqrt{484}}{2} = \frac{-8 \pm 22}{2}.$$

This quadratic equation has two solutions, $y = -15$ or $y = 7$. Substituting y into (1), we get two quadratic equations in x,

$$x^2 + 6x + 15 = 0 \text{ and } x^2 + 6x - 7 = 0.$$

Consider first the equation $x^2 + 6x + 15 = 0$.

$$D = 6^2 - 4 \cdot 15 = 36 - 60 = -24 < 0.$$

Since the discriminant is a negative number, the first equation has no real solutions.

Consider now the second equation $x^2 + 6x - 7 = 0$.

$$D = 6^2 + 4 \cdot 7 = 64.$$

$$x = \frac{-6 \pm \sqrt{64}}{2} = \frac{-6 \pm 8}{2}.$$

There are two solutions of the second equation, $x = 1$ or $x = -7$. Only the positive number satisfies the conditions of the problem.

Therefore,

$$a_1 = 1,$$

$$a_2 = x + 2 = 1 + 2 = 3,$$

$$a_3 = x + 4 = 1 + 4 = 5,$$

$$a_4 = x + 6 = 1 + 6 = 7.$$

You'll certainly enjoy investigating different solutions to the above problems. I believe you will then truly appreciate the introduced formulas for their efficiency and effectiveness.

Heron's formula

$$S = \sqrt{p(p-a)(p-b)(p-c)}, \text{ where } p = \frac{a+b+c}{2}.$$

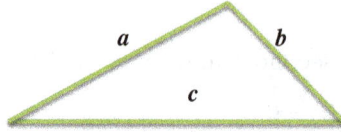

Brahmagupta's formula

$$S = \sqrt{(p-a)(p-b)(p-c)(p-d)}$$

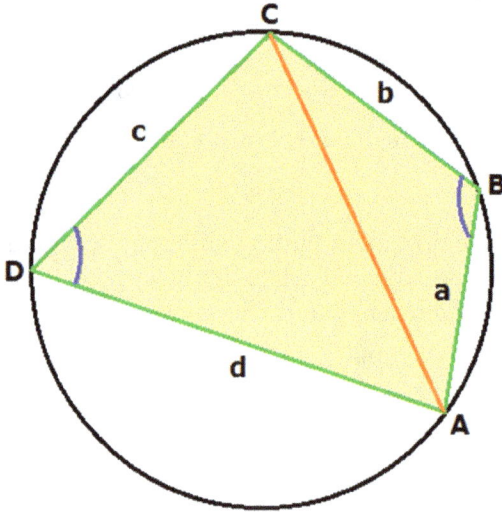

$$p = \frac{a+b+c+d}{2} \qquad\qquad \angle B + \angle D = 180°$$

Chapter 6

The Theorem of Ratios of the Areas of Similar Polygons

When I was playing with the idea of putting together my published articles in various math magazines into this book, I never intended to make it a routine set of problems or a textbook. The main idea was to have an emphasis on various problem-solving methods and techniques, and to explore little-known or unfamiliar facts from plane geometry. We concentrate on some non-standard geometrical problem-solving techniques and analyze how to solve difficult problems in the most efficient way and how to get the solutions to look elegant, non-threatening, and relevant.

The Theorem of Ratios of the Areas of Similar Polygons is a powerful and useful tool in simplifying solutions of many geometric problems involving similar polygons.

In real life, geometry has many practical uses, from the most basic to those in the most advanced phenomena in our life. Geometry plays an especially essential role in various construction and engineering projects. "For without symmetry and proportion no temple can have a regular plan", stated the ancient Roman architect Marcus Vitruvius in his famous book *De Architectura*. Area problems involving similar polygons commonly occur in architectural applications.

The theorem in this chapter connects the areas of similar polygons with their corresponding linear elements. It incorporates the application of a "theoretical" concept to real-life situations.

Similar polygons in real life

By applying the methods below, the reader should be able to obtain elegant solutions to certain problems posed in previous chapters as well.

Let's proceed to the theorem:

Theorem. *The ratio of the areas of the two similar polygons is equal to the squared ratio of their corresponding linear elements.*

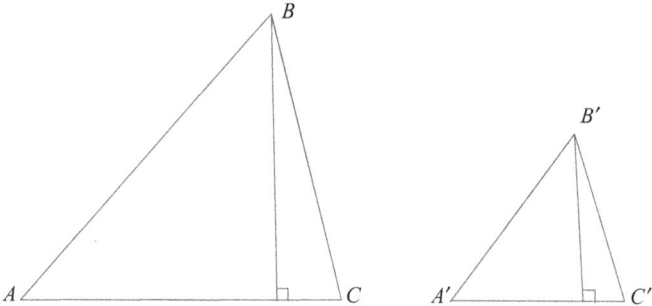

Proof. The theorem can be proved by induction.

First, we consider the pair of similar triangles ABC and $A'B'C'$.

The area of the triangle ABC, $S = \frac{1}{2}ah$, and the area of the triangle $A'B'C'$, $S' = \frac{1}{2}a'h'$, where a, h, a', and h' are the corresponding bases and altitudes in the triangles ABC and $A'B'C'$ respectively.

Since the triangles are similar, the ratio of their linear elements (corresponding sides, altitudes, etc.) will always be the same. If we denote by x the ratio of the corresponding linear elements of these triangles, we obtain

$$\frac{S}{S'} = \frac{\frac{1}{2}ah}{\frac{1}{2}a'h'} = \frac{a}{a'} \cdot \frac{h}{h'} = x \cdot x = x^2.$$

It is important to bear in mind this fact, because we will use it in some following problems as well.

Now, let's take two similar n-gons. Each of them can be partitioned into $n - 2$ similar triangles (by diagonals drawn from the corresponding vertices, for example). See figure below.

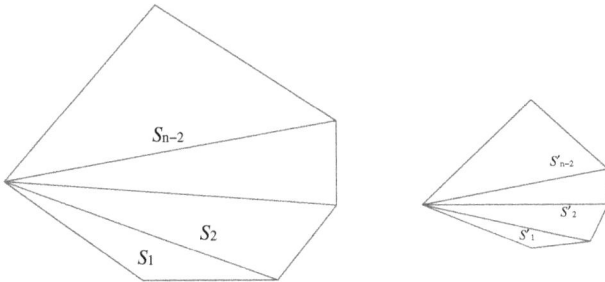

To make it easier to understand the following argument, we will first consider the sum of the areas of two such triangles:

$$S_1 + S_2 = \frac{1}{2}(a_1 h_1 + a_2 h_2) = x^2 \frac{1}{2}(a_1' h_1' + a_2' h_2') = x^2(S_1' + S_2').$$

Next, assuming the equality has already been proved for $n = k - 1$, let's make sure it's true for $n = k$.

Using the induction assumption and the fact that for the similar triangles the equality has already been proved, we get

$$S = S_1 + S_2 + \cdots + S_{k-2} = (S_1 + S_2 + \cdots + S_{k-3}) + S_{k-2}$$

$$= x^2 \cdot \sum_{i=1}^{k-3} S_i' + x^2 \cdot S_{k-2}' = x^2 \left(\sum_{i=1}^{k-3} S_i' + S_{k-2}' \right) = x^2 \cdot S',$$

where $S_i (i = 1, 2, \ldots, k - 2)$ and $S_i' (i = 1, 2, \ldots, k - 2)$ represent the areas of the corresponding similar triangles into which each n-gon has been partitioned. The proof is complete.

Now we are going to demonstrate the theorem's application in problem solving.

Problem 1. In triangle ABC, points A' and C' are the bases of the altitudes dropped to sides BC and AB respectively. Prove that if $\angle ABC = 60°$, then the area of the triangle $A'BC'$ is 25% of the area of the triangle ABC.

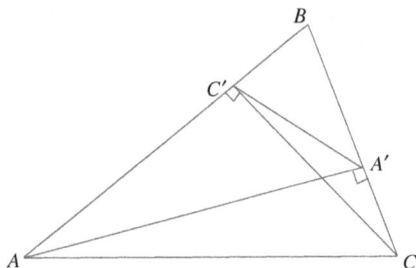

Proof.   Considering the triangles $A'BC'$ and ABC, we know that they are similar, with the ratio $k = \cos B$ (see Theorem 1 from the chapter "The Orthic Triangle and Some of Its Properties"). If we denote by S the area of the triangle ABC and by S' the area of the triangle $A'BC'$, then $S : S' = k^2 = \cos^2 60° = \frac{1}{4}$, which was to be proved.

Problem 2. Three lines parallel to the sides of the triangle ABC are drawn through the point M lying inside the triangle. The lines partitioned $\triangle ABC$ into six parts, three of which are triangles with areas S_1, S_2, and S_3. Find the area of the triangle ABC.

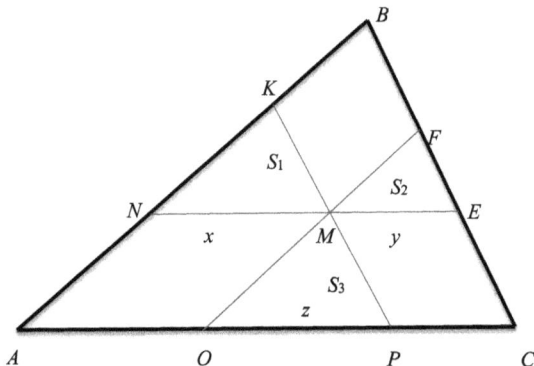

Solution. By the conditions of the problem, we know that $NE \parallel AC$, $OF \parallel AB$, and $KP \parallel BC$. Then obviously, the triangles NKM, MFE, and OMP are similar to each other and to triangle ABC because they have the same three angles. Denote the area of triangle ABC by S, the area of triangle NKM by S_1, the area of triangle MFE by S_2, and the area of triangle OMP by S_3, and let $MN = x$, $ME = y$, $OP = z$.

Applying the theorem to pairs of similar triangles, we have

for $\triangle ABC$ and $\triangle NKM$, $\sqrt{S_1} : \sqrt{S} = x : AC$,

for $\triangle ABC$ and $\triangle MFE$, $\sqrt{S_2} : \sqrt{S} = y : AC$,

for $\triangle ABC$ and $\triangle OMP$, $\sqrt{S_3} : \sqrt{S} = z : AC$.

Adding, we get

$$\frac{\sqrt{S_1} + \sqrt{S_2} + \sqrt{S_3}}{\sqrt{S}} = \frac{x + y + z}{AC}. \qquad (*)$$

Note that $ANMO$, $MKBF$, and $MECP$ are the parallelograms (as given, in each quadrilateral the opposite sides are parallel). Then $x = NM = AO$ and $y = ME = PC$. It's easy to see that $x + y + z = AO + OP + PC = AC$. By substituting this into $(*)$, we conclude that

$$\frac{\sqrt{S_1} + \sqrt{S_2} + \sqrt{S_3}}{\sqrt{S}} = 1,$$

from which we arrive at

$$S = (\sqrt{S_1} + \sqrt{S_2} + \sqrt{S_3})^2.$$

Problem 3. CD is an altitude in the right triangle $ABC(\angle C = 90°)$, r' and r'' are the radii of the inscribed circles in the triangles ADC and BDC respectively. Find the radius of the circle inscribed in the triangle ABC.

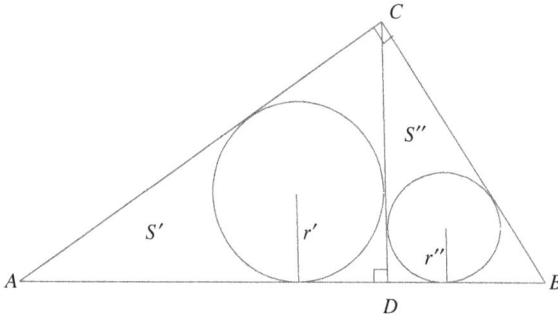

Solution. It's noteworthy that it is always up to us how to select linear elements in similar triangles for the best application of the

Theorem of the ratios of the areas of similar polygons. In this case, we will work with the radii of inscribed circles in similar triangles.

Right triangles ADC and ACB, and BDC and BCA are similar because they share a common angle A and common angle B respectively. Denoting the area of the triangle ABC by S, the area of the triangle ADC by S', the area of the triangle BDC by S'', and the radius of the inscribed circle by r, we obtain

$$\frac{S'}{S} = \frac{r'^2}{r^2}, \quad \frac{S''}{S} = \frac{r''^2}{r^2}.$$

Adding the equalities yields

$$\frac{S' + S''}{S} = \frac{r'^2 + r''^2}{r^2}.$$

Because $S' + S'' = S$, we get $\frac{r'^2 + r''^2}{r^2} = 1$, from which $r^2 = r'^2 + r''^2$, and we are done.

Expressing the obtained result in words, we arrive at the following nice property of the right triangle:

The square of the radius of the inscribed circle in the right triangle is the sum of the squares of the radii of two circles inscribed in triangles partitioned from the original triangle by its altitude dropped to the hypotenuse.

Problem 4. In trapezoid $ABCD$ ($BC \parallel AD$), O is the point of intersection of its diagonals. The area of the triangle BOC is S', and the area of the triangle AOD is S''. Find the area of $ABCD$.

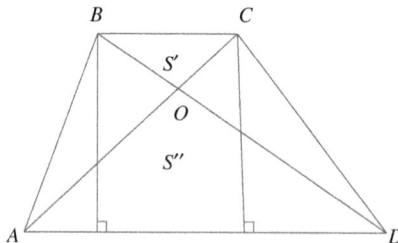

Solution. We preface the solution of this problem with the important observation:

In trapezoid $ABCD$ ($BC \parallel AD$), the areas of the triangles AOB and COD are equal.

This assertion can easily be proved by comparing the areas of triangles ABD and ACD.

Triangles ABD and ACD have the common base AD and equal altitudes dropped to it. Therefore, their areas are equal. Each of the triangles ABD and ACD consists of a common part, triangle AOD, and the triangles AOB and DOC respectively. Then the area of AOB must equal to the area of DOC, as we wanted to prove.

We will now return to our original problem.

Our goal will be to express the area of $\triangle AOB$ or $\triangle DOC$ in terms of S' and S''. If we manage to do it, the last step in finding the area of $ABCD$ will be just to add the areas of the triangles AOB, BOC, DOC, and DOA.

From the similar triangles BOC and DOA (respective angles are congruent), we get $\frac{CO}{OA} = \frac{\sqrt{S'}}{\sqrt{S''}}$, or expressing OA in terms of CO,

$$OA = CO \cdot \frac{\sqrt{S''}}{\sqrt{S'}}. \tag{1}$$

The area of the triangle AOB is

$$S_{AOB} = \frac{1}{2} \cdot BO \cdot OA \cdot \sin \angle BOA.$$

Substituting OA from (1) we get

$$S_{AOB} = \frac{1}{2} \cdot BO \cdot CO \cdot \frac{\sqrt{S''}}{\sqrt{S'}} \cdot \sin \angle BOA = \left(\frac{1}{2} \cdot BO \cdot CO \cdot \sin \angle BOA \right) \cdot \frac{\sqrt{S''}}{\sqrt{S'}}. \tag{2}$$

Now, observe that $\angle BOA = 180° - \angle BOC$ and $\sin \angle BOA = \sin(180° - \angle BOC) = \sin \angle BOC$.

Also, S', the area of $\triangle BOC$, can be calculated as $S' = \frac{1}{2} \cdot BO \cdot CO \cdot \sin \angle BOC$.

Substituting $\sin \angle BOC$ for $\sin \angle BOA$ into (2), we obtain

$$S_{AOB} = \left(\frac{1}{2} \cdot BO \cdot CO \cdot \sin \angle BOC \right) \cdot \frac{\sqrt{S''}}{\sqrt{S'}} = S' \cdot \frac{\sqrt{S''}}{\sqrt{S'}} = \sqrt{S'S''}.$$

To summarize, we just got that

$$S_{AOB} = S_{DOC} = \sqrt{S'S''}.$$

Finally, to find the area of the trapezoid, we have to add the areas of the four triangles into which it is partitioned by its diagonals:

$$S = S' + S'' + 2\sqrt{S'S''} = (\sqrt{S'} + \sqrt{S''})^2.$$

Problem 5. In trapezoid $ABCD$ ($AB \parallel CD$), $AB = a$ and $CD = b$. MN is parallel to the bases of the trapezoid and separates it into two trapezoids of equal area. Find the length of MN.

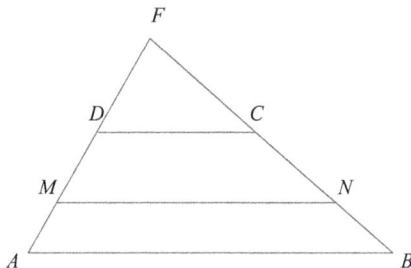

Solution. Extend the sides AD and BC to their intersection at F. The newly formed triangles AFB, MFN, and DFC share angle F. Also, as given, $MN \parallel AB \parallel DC$, which implies the congruence of the respective angles by the parallel bases in these triangles. Thus, triangles AFB, MFN, and DFC are similar, since their corresponding angles are equal.

For convenience, denote $MN = x$ and $S_{AMNB} = S_{MDCN} = S$.

From the similarity of the triangles AFB and MFN, it follows that

$$S_{AFB}/S_{MFN} = (a/x)^2. \tag{1}$$

From the similarity of the triangles MFN and DFC, it follows that

$$S_{DFC}/S_{MFN} = (b/x)^2. \tag{2}$$

The triangle AFB consists of the triangle MFN and the trapezoid $AMNB$. Its area is

$$S_{AFB} = S_{MFN} + S. \tag{3}$$

Analogously, triangle MFN consists of the triangle DFC and the trapezoid $MDCN$. Therefore,

$$S_{MFN} = S_{DFC} + S,$$

from which

$$S_{DFC} = S_{MFN} - S. \tag{4}$$

Substituting the areas of the triangles AFB and DFC from (3) and (4) respectively into (1) and (2), we get
$S_{AFB}/S_{MFN} = (S_{MFN} + S)/S_{MFN} = 1 + S/S_{MFN} = (a/x)^2$, from which

$$S/S_{MFN} = (a/x)^2 - 1. \tag{5}$$

$$S_{DFC}/S_{MFN} = (S_{MFN} - S)/S_{MFN} = 1 - S/S_{MFN} = (b/x)^2,$$

from which

$$S/S_{MFN} = 1 - (b/x)^2. \tag{6}$$

Comparing (5) and (6) yields an equation $(a/x)^2 - 1 = 1 - (b/x)^2$. After a few simplifications, we get that $x = \sqrt{\frac{a^2+b^2}{2}}$.

The solution is not unique. I encourage readers to look for alternative solutions. For example, try to work with the areas of the three trapezoids $AMNB$, $MDCN$, and $ADCB$ and express the length of MN using the formula for the area of a trapezoid.

However, be careful when identifying similar figures. In this problem, I saw students assume that the trapezoids $AMNB$ and $MDCN$ are similar. Such an assumption led to a wrong result. It's critical to prove the similarity of the figures you work with before you apply the Theorem of the areas of similar polygons to them.

In conclusion, here are a few problems for you to solve:

Problem 6. In triangle ABC, K is a random point lying on AC. Draw the straight line at K such that it separates the triangle ABC into two parts of equal area.

Problem 7. Let D be a point on the side AB of the triangle ABC. The lines through D parallel to the sides AC and BC intersect them at F and E, respectively. Find the area of the triangle CEF, given that the area of the triangle ADF is S'' and the area of the triangle BDE is S'.

Problem 8. Prove that the area of the triangle with sides formed by the midlines of a triangle equals $\frac{1}{4}$ of the area of that triangle.

Problem 9. Prove that the area of the triangle with sides formed by the medians of another triangle equals $\frac{3}{4}$ of the area of that triangle.

The next problem is a strengthening of the famous Pythagorean Theorem. The statement of this problem seems to have many interesting consequences. For example, earlier solved Problem 3 is a direct corollary of it. This property is often useful in solving problems concerning a right triangle with the altitude drawn to the hypotenuse.

Problem 10. In a right triangle, the altitude drawn to the hypotenuse divides it into two triangles such that for any corresponding linear elements l_1, l_2, and l of the two triangles and the original triangle, $l^2 = l_1^2 + l_2^2$.
See figure below:

l_1 is a linear element of triangle CDB,
l_2 is a linear element of triangle ADC,
l is a linear element of triangle ACB.

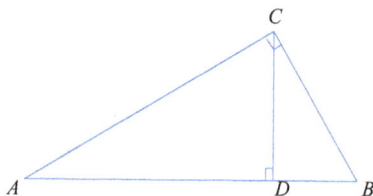

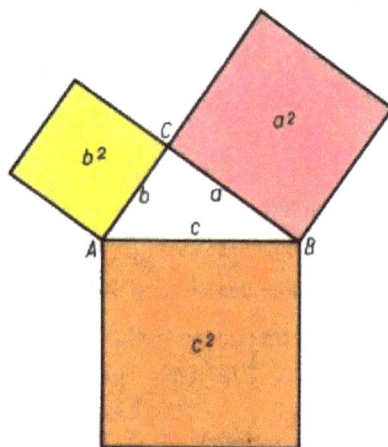

Chapter 7

A Pivotal Approach: Applying Rotation in Problem Solving

In previous chapters, we used line and point reflections to solve several problems. The properties of line symmetry and point symmetry might significantly simplify the solution and provide elegant and effective ways to the final result. In this chapter, we will use one more important kind of transformation that preserves the distance — rotation.

By the definition of rotation, the entire plane is turned about some point through an angle clockwise or counterclockwise. Thus the size and shape of any figure are kept invariant, but its points all move along arcs of concentric circles. The center (which may or may not belong to the figure being rotated) is the only point that remains fixed.

Because rotation preserves distance, it takes any figure into a congruent figure. The angles between corresponding lines are equal to each other and to the given angle of rotation. These very important properties of rotation can be widely used in problem solving.

Here we'll examine some methods and techniques that use rotation and its properties.

To start, let's perform a simple rotation of a segment AB about a point O (the center) through an angle θ (see figure below). AB is transformed into $A'B'$. The angle between the lines containing the segments equals θ, and by properties of rotation, $AB = A'B'$.

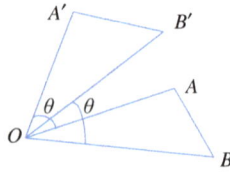

All the constructions that follow are based on these results, and all the proofs are based on the same properties of rotation.

Problem 1. Two squares $AMKB$ and $ACPT$ are drawn externally on the sides AB and AC of a triangle ABC. Prove that the distance between points M and T is equal to twice the length of the median of triangle ABC drawn to side BC.

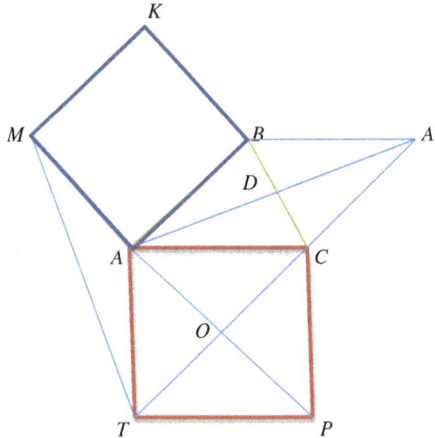

Solution. Denote by O the point of intersection of the diagonals of the square $ACPT$. Consider the 90° rotation of the square about its center O, taking T into A and A into C (it's a clockwise rotation). Before going on, we make one auxiliary construction: draw a line through B parallel to AC and draw a line through C parallel to AB. Let A' be their point of intersection and let D be the midpoint of BC. Thus we obtained a parallelogram $ABA'C$, whose center is D. Let's show that the rotation takes the segment MT into AA' — this will prove the statement of the problem.

Since $MA = AB$ and $AB = A'C$, then $MA = A'C$. We know that $MA \perp AB$ and $AB \parallel A'C$; therefore, $MA \perp A'C$. This means that the segment MA is rotated into the segment $A'C$, so M is rotated into A'. We have already noted that T is rotated into A. Thus, the segment MT is rotated into the segment AA', and therefore, $MT = AA'$ by the properties of rotation. The segment AD is the median of triangle ABC and $AD = \frac{1}{2}AA'$ (by the property of the parallelogram); hence, $MT = 2AD$, and the proof is complete.

Problem 2. The points M and N are chosen on the sides BC and CD, respectively, of a square $ABCD$ such that $BM{:}MC = 3{:}1$ and $CN{:}ND = 3{:}1$. Prove that $AM{\perp}BN$.

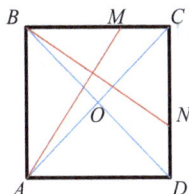

Proof. Let O be the center of the square. A $90°$ rotation about O clockwise takes B into C, and C into D. Since our rotation takes BC into CD, and since $BM{:}MC = CN{:}ND = 3{:}1$, then M must be taken into the point that divides CD in the same ratio, as M divides BC, that is, M rotates into N. According to properties of rotation, the angle between corresponding rays must be $90°$, therefore, AM is perpendicular to BN, which was to be proved.

An interesting generalization follows:

If there are two points M and N selected on the adjacent sides of a square $ABCD$ (M on BC and N on CD), such that they split the sides in the same ratio, then the segments BN and AM will be perpendicular no matter what the ratio is.

Problem 3. Construct an equilateral triangle such that its vertices are located on the three given concentric circles.

Remember, all constructions in this book are to be performed only by means of compass and straightedge, unless indicated otherwise.

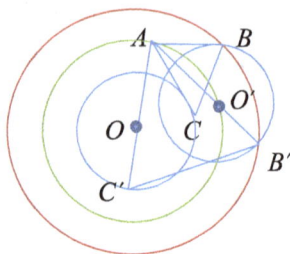

Solution. Choose any point A on the middle circle and rotate the smallest circle about A through an angle of 60°. The circle is transformed into a circle with center O' with the same radius. Denote the points of its intersection with the biggest circle by B and B'. (If there is only one point of intersection, the problem has a unique solution; if the circles do not intersect, the problem has no solution.) The last steps are to draw the circle with center at B and radius AB, and the circle with center at B' and radius AB'. The points of intersection of these circles with the smallest circle give us the third vertices of the triangles ABC and $AB'C'$ — the desired equilateral triangles with vertices on the three given concentric circles. It would be good practice for readers to prove that the triangles satisfy the conditions of the problem.

The rotation that has the most interesting properties of all is the one that transforms every line into a line parallel to it. This is the *half-turn*, or rotation through 180°, which transforms each ray into an oppositely directed ray. Clearly, a half-turn is completely determined by its center. Another name for the half-turn is *central symmetry*. A point and its image are called symmetric with respect to the center of rotation.

Problem 4. Given an angle and a point M inside it, draw a segment with its midpoint at M and its endpoints on the sides of the angle.

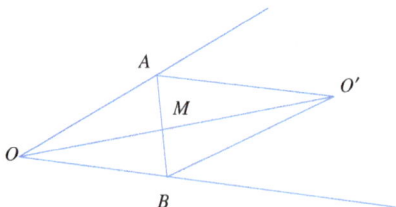

Solution. Let O be the vertex of the angle. The half-turn about M takes O into O' such that $OM = O'M$ and all three points lie on one line. Draw lines through O' parallel to the sides of the angle. Denote the points of intersection of these lines with the angle's sides by A and B. Then $OAO'B$ is a parallelogram by construction. Since M is the midpoint of OO' (by the property of half-turn), it must be the midpoint of the second diagonal of the parallelogram $OAO'B$ as well. Therefore, AB is the solution. The reader may verify that there are no more solutions.

Problem 5. The midpoint M of the side AB of a trapezoid $ABCD$ ($BC \parallel AD$) is connected to points C and D. Prove that the area of the triangle MCD is $\frac{1}{2}$ of the area of trapezoid $ABCD$.

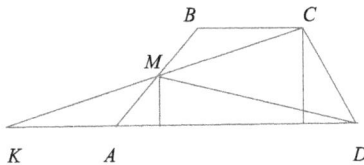

Solution. To solve the problem, we first find the point K that is symmetrical to C with respect to M. We will show that K necessarily lies on the extension of the side AD.

Since $BM = MA$ (by the conditions of the problem), $MC = MK$, and $\angle BMC = \angle AMK$ (by construction), triangles BMC and AMK are congruent by SAS property. Therefore, $\angle MKA = \angle MCB$ as respective angles in congruent triangles. But these angles are alternate interior angles formed by transversal KC with BC and AK. Hence, AK must be parallel to BC, which means that K is on the extension of AD. Being congruent, triangles AMK and BMC have equal areas. Thus, the area of the trapezoid $ABCD$ is equal to the area of triangle KCD (they both consist of common part $AMCD$ and congruent triangles AMK and BMC). To complete the proof, it suffices to show that the area of KCD is equal to twice the area of triangle MCD. This is easy — DM is a median in triangle KDC, and a median always divides a triangle into two triangles of equal area (see Chapter 1).

Problem 6. Given three points O, M, and N, construct a square such that O is its center and points M and N lie on opposite sides of the square (or their extensions).

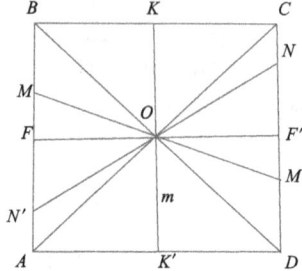

Solution. Construct M' symmetric to M and N' symmetric to N with respect to O. Then the lines MN' and NM' are parallel by the properties of central symmetry. Next, draw the perpendicular FF' through O to lines MN' and $M'N$ (F lies on MN' and F' lies on NM') and the line m passing through O parallel to MN' and $M'N$. Locate the points K and K' on m such that $OK = OK' = OF = OF'$. To complete the construction, draw lines through K and K' parallel to FF'. Denote the points of intersection of these lines with MN' and NM' as B, A, D, and C respectively. $ABCD$ is a rectangle, and $AB = BC = CD = AD$ (by construction). Therefore, $ABCD$ is the desired square.

Sometimes M and N lie on the sides of a square, and sometimes on their extensions. The reader can investigate different cases. What happens if M and N are reflections of each other in O?

"The advancement and perfection of mathematics are intimately connected to the prosperity of state".

Napoleon Bonaparte

These words belong to French Emperor Napoleon Bonaparte (1769–1821), who always expressed great interest in mathematics. He kept close connections with some of the most outstanding mathematicians of his time, including Gaspard Monge, Pierre Simon Laplace, Lorenzo Mascheroni, Adrien-Marie Legendre, and Joseph Louis Lagrange.

One of the classic historic problems in plane geometry bears his name and is known as *Napoleon's Theorem:*

If equilateral triangles are erected externally on the sides of any triangle, their centers form an equilateral triangle.

It was speculated that this problem was mentioned in one of the letters by Napoleon Bonaparte to the prominent mathematician, member of the French Academy, and first professor of analysis at the Ecole Polytechnique, Joseph Louis Lagrange (1736–1813).

Was Napoleon really the first to discover the proof? From what I read, there is no clear evidence of that fact. It is possible, but does it really make any difference? The problem is remembered by his name and we recall it here as a good example of applying rotation in problem solving.

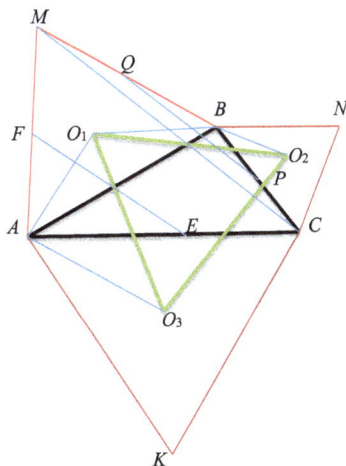

Solution. ABC is the given triangle; AMB, BNC, and ACK are the equilateral triangles constructed on its sides. Denote the centers of the equilateral triangles by O_1, O_2, and O_3. We need to prove that the triangle $O_1O_2O_3$ is equilateral.

To do it, we consider two different rotations. First, we rotate O_1O_3 counterclockwise through an angle of 30° about A. Then O_1 will be rotated into F on AM such that $FA = AO_1$ (O_1 is the center of the equilateral triangle AMB, therefore $\angle FAO_1 = 30°$). Also, O_3 will be rotated into E on AC such that $AE = AO_3$ (O_3 is the center of the equilateral triangle AKC, therefore $\angle EAO_3 = 30°$). FE is the image of O_1O_3, which means that $FE = O_1O_3$.

Second, we rotate O_1O_2 clockwise through an angle of 30° about B. Then O_1 is rotated into Q on BM such that $BQ = BO_1$ ($\angle O_1BQ = 30°$) and O_2 is rotated into P on BC such that $BP = BO_2(\angle O_2BP = 30°)$.

Let's take a small step aside and quickly prove one useful formula for an equilateral triangle relationship:

In an equilateral triangle, the distance from a vertex to the center is $\frac{\sqrt{3}}{3}$ of its side.

Denote a the length of the side in an equilateral triangle XYZ, $XY = YZ = XZ = a$. Draw the altitudes (in equilateral triangle, they are the medians and angle bisectors at the same time) YH and XG and denote O the point of their intersection.

Applying the Pythagorean Theorem to right triangle XHY ($\angle H = 90°$) we have

$$YH = \sqrt{a^2 - \left(\frac{a}{2}\right)^2} = \sqrt{\frac{3a^2}{4}} = \frac{a\sqrt{3}}{2}.$$

Recalling that O is the centroid of $\triangle XYZ$ and that it is $\frac{2}{3}$ of the way from a vertex to the opposite midpoint, we arrive at $OY = \frac{2}{3} \cdot \frac{a\sqrt{3}}{2} = \frac{a\sqrt{3}}{3}$, which is the relationship we set to develop.

Turning back to our original problem, in the equilateral triangles AMB, AKC, and BNC we see that

$$AO_1/AM = AF/AM = \frac{\sqrt{3}}{3},$$

$$AO_3/AC = AE/AC = \frac{\sqrt{3}}{3},$$

$$BO_1/BM = BQ/BM = \frac{\sqrt{3}}{3}, \text{ and}$$

$$BO_2/BC = BP/BC = \frac{\sqrt{3}}{3}.$$

It follows that the pairs of triangles FAE and MAC, and QBP and MBC are similar (because of the two proportional sides and common angle between them) with the same ratio $\frac{\sqrt{3}}{3}$, implying that $FE \parallel MC \parallel QP$ and $FE = QP$ (because $FE = \frac{\sqrt{3}}{3}MC$ from the similarity of triangles FAE and MAC and also, $QP = \frac{\sqrt{3}}{3}MC$ from the similarity of triangles QBP and MBC).

Now we recall that $FE = O_1O_3$ as images in the first rotation and $QP = O_1O_2$ as images in the second rotation.

Therefore, $O_1O_3 = O_1O_2$.

Similarly, $O_2O_3 = O_1O_3 = O_1O_2$ and the proof is completed. As we wanted to show, triangle $O_1O_2O_3$ is equilateral.

The solution of course is not unique, but perhaps it is the simplest one. It provides a great demonstration of applying rotation. You might want to explore the other solutions to this great problem. To name a few, the composition of a few rotations by $120°$ with the center of rotation being one of the centers of the equilateral triangles, an application of the Law of cosines and trigonometric manipulations, a proof based on complex number arithmetic, a pure geometric proof based on a property of circumscribed circles. It is noteworthy that this remarkable problem relates to the famous Fermat point of a triangle: a point with the minimal total distance to the vertices. The Fermat point gives a solution to the geometric median and Steiner tree problems for three points. Napoleon's Theorem has a few generalizations, which we are not going to mention here. You can go on and on and on ... It might become a subject matter of a new book. Is not it amazing how one historic problem gives birth to an expanded family of relatives?

To sharpen your skills in applying rotation, we leave it to you to solve the following construction problems.

Problem 7. There is given an acute angle with vertex A and arbitrary point O interior to it. Construct a straight line passing through O, which cuts from the angle the triangle of minimal area.

Problem 8. M is an arbitrary point on the side AB of a square $ABCD$. Construct a square inscribed in $ABCD$ such that M is one of its vertices.

Problem 9. Construct an equilateral triangle such that its three vertices lie on three given parallel lines.

Chapter 8

Auxiliary Elements in Problem Solving

One day when I was leafing through an old July/August 1996 issue of the currently defunct magazine *Quantum*, I was stimulated by the list of problems "What's the "best" answer?" by Boris Kordemsky. The author considered several interesting algebra problems and proposed "... short and ... beautiful solutions involving some witty and useful tricks." In the solutions of Problems 11, 12, and 14 the "trick" was using the method of an auxiliary element — a powerful and useful tool in problem-solving.

I would recommend readers to return to the problems mentioned below. Try to solve them after reading this chapter (even though these problems have nothing to do with geometry, solving them will provide a better understanding of the auxiliary element concepts covered in this chapter).

Problem 11. Simplify the expression
$$\frac{(x-a)(x-b)}{(c-a)(c-b)} + \frac{(x-b)(x-c)}{(a-b)(a-c)} + \frac{(x-c)(x-a)}{(b-c)(b-a)}.$$

Problem 12. Solve the equation $x^3 + 1 = 2\sqrt[3]{2x-1}$.

Problem 14. Simplify the expression $\sin^3 x \cos 3x + \cos^3 x + \sin 3x$.

In Problem 11, the expression was simplified by introducing an auxiliary function — a polynomial of degree no greater than 2. In solving equation 12, the author made use of an auxiliary function

$f(x) = \sqrt[3]{2x - 1}$. In Problem 14, the author again introduced an auxiliary function, the derivative of which helped make the solution short and elegant.

The concept of an auxiliary element can be widely used not only in algebra and calculus, but in geometry as well.

In simple words, we can define problem solving as a process of building a logical chain by finding connections between given elements. The shorter the chain is the better. We are always interested in the optimal solution. That's why the method of auxiliary elements is so important — generally it gives us the shortest and the most elegant solution of a problem. One of the most important strategic components of mathematical power is the ability to evaluate a problem — to determine from the first steps, the most important properties of the given elements and how to connect them. When you face more sophisticated problems, for instance, like those mentioned above, when you do not see on the surface connections between elements, you might give up, unable to pursue a different line of reasoning. An auxiliary element enables us to clarify the picture by making use of the properties of a new element, and simplifying the process of finding the ties and connections that were hidden before. This is another reason why I think the introducing auxiliary elements is so useful and that is why I will examine it in this chapter.

Let us start with a challenging problem that was on one of the math contests many years ago:

Karabases and Barabases are the two biggest ethnic groups among the citizens of a mysterious country. Each Karabas has 11 friends that are Karabases and 8 friends that are Barabases, each Barabas has 9 friends that are Barabases and 10 friends that are Karabases. What number is greater: the number of Karabases or the number of Barabases?

This nonstandard, cute problem turned out to be a real disaster in the contest. None of the students provided a reasonable and clear explanation of the comparison of Karabases and Barabases. Before you go on, try to solve the problem on your own. Give yourself a few minutes (if necessary, hours) to think it over. You will get real pleasure when you find the answer, if, of course, you manage to solve the problem. In approaching the solution to such a puzzle like this, it is helpful to make use of the method of the auxiliary element. The "Karabases-Barabases" problem in my opinion is an excellent, vivid example of the concept. The difficulty is to establish any kind of a relationship between Karabases and Barabases. Really, you have no idea what the number of Karabases is, or what the number of Barabases is. So, how can you compare unknown, and even worse, undeterminable numbers? But remember, we are not asked to find the number of Karabases and Barabases, we are asked to *compare* them. The clue to the solution lies in this word — "compare". We should somehow find an expression connecting the number of Karabases and Barabases that will allow us to compare them. Here the role of the auxiliary element is played by an ordinary business card. Just assume that each Karabas gives his business card to each of his Barabases friends. If we denote the number of Karabases by K and the number of Barabases by B, then the total number of cards given equals $8B$ (since each Karabas has 8 friends among the Barabases). On the other hand, the total number of cards received by Barabases equals $10K$ (each Barabas gets a business card from his 10 Karabases friends). Making use of such an auxiliary element as a business card allows us to set up a simple equality $8B = 10K$, which can be true only when $B > K$, or, in words, the number of Barabases must be greater than the number of Karabases. Is it not fascinating how simple and elegant the solution turned out to be? By the way, notice that the number of the Karabases-friends of the Karabases and the number of the Barabases-friends of the Barabases were not used at all in the problem. Those conditions just add confusion, but play no role whatsoever in achieving the final result.

In this chapter, I'd like to propose several challenging and interesting problems in plane geometry. Their solutions obtained by introducing such auxiliary elements as circles, triangles, trapezoids, etc., are attractive not only because of their elegance, but their effectiveness as well. One can raise a number of questions: How did you manage to find this particular auxiliary element? Why did you introduce it? What is a pattern or clue to the selection of a specific auxiliary element? Well, we will try our best to give reasonable answers to all of the questions in the problems considered below. But before going on, I would like to emphasize one more time that it is critical to evaluate the most essential properties of the given elements and take advantage of them when deciding which auxiliary elements to introduce.

Problem 1. Find the sum of squares of the medians of a triangle if its sides are a, b, c.

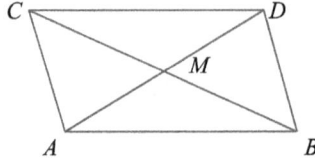

Solution.

To solve the problem, we need to express each median in terms of the sides of the triangle and then to add the expressions. In this case, it would be useful to construct an auxiliary parallelogram. Why a parallelogram and not something else? Well, take a careful look at the diagram — the median bisects opposite side, so if we extend AM beyond M and locate D on the extension of AM such that $DM = MA$, then in the quadrilateral $ABDC$ diagonals AD and BC bisect each other. Therefore, $ABDC$ is a parallelogram. And now our problem transforms into another problem: express a diagonal of a parallelogram in terms of its sides and the other diagonal.

In some cases, to solve a complex problem, you may dissect it into a few small problems. Taking advantage of the results from these auxiliary problems one can significantly simplify the path to the solution of the original one. Here we will take a step aside and will prove a very useful property of a parallelogram (we will apply it to solve our problem) that the sum of squares of its diagonals equals the sum of squares of its sides.

Applying the Law of cosines to $\triangle AMC$, we have

$$AC^2 = AM^2 + CM^2 - 2AM \cdot CM \cos \angle AMC. \qquad (1)$$

Similarly, Applying the Law of cosines to $\triangle CMD$ and recalling that $\cos(180° - \angle AMC) = -\cos \angle AMC$, we get

$$CD^2 = DM^2 + CM^2 - 2DM \cdot CM \cos \angle DMC$$
$$= DM^2 + CM^2 - 2DM \cdot CM \cos(180° - \angle AMC)$$
$$= DM^2 + CM^2 + 2DM \cdot CM \cos \angle AMC. \qquad (2)$$

In parallelogram $ABDC$, $AC = BD$ and $AB = CD$ as opposite sides. Also, $AM = MD$ and $CM = MB$ because the point of intersection of the diagonals bisects them.

Adding (1) and (2) leads to

$AC^2 + CD^2 = AM^2 + CM^2 - 2AM \cdot CM \cos \angle AMC + DM^2 + CM^2 + 2DM \cdot CM \cos \angle AMC = 2AM^2 + 2CM^2 + 2CM(DM - AM) \cos \angle AMC = 2AM^2 + 2CM^2$. Multiplying by 2 both sides of the last equality and substituting $2AM$ for AD and $2CM$ for CB we arrive at the relationship we set to develop:

$$2AC^2 + 2CD^2 = AD^2 + CB^2. \tag{3}$$

We will now return to our original problem.

We know that $BC = a$, $AC = b$, and $AB = c$, and if we denote AM by x, $AM = x$, then $AD = 2x$. In our nominations (3) can be written as $(2x)^2 + a^2 = 2b^2 + 2c^2$. After simplification, we obtain that $x^2 = \frac{1}{4}(2b^2 + 2c^2 - a^2)$. Denoting the medians to sides AB and AC by y and z respectively, and by making use of two more auxiliary parallelograms constructed in a similar way, we get $y^2 = \frac{1}{4}(2a^2 + 2b^2 - c^2)$ and $z^2 = \frac{1}{4}(2a^2 + 2c^2 - b^2)$. Adding and simplifying, we obtain the desired result, $x^2 + y^2 + z^2 = \frac{3}{4}(a^2 + b^2 + c^2)$.

I'd like to emphasize that the transformation of the problem from the triangle to an auxiliary parallelogram helped us to establish the relationship between the given elements and the unknown medians pretty easily. The decision to use a parallelogram as an auxiliary element was not so obvious. It required certain imagination and creativity. But as a matter of fact, it was a logical conclusion from the definition of a median of a triangle. In many problems, when you deal with the relationships between the sides and a median of a triangle, the above trick with the construction of an auxiliary parallelogram might be helpful.

Problem 2. ABC is an equilateral triangle. A ray is drawn from A, and M is picked on it such that $\angle BMA = 20°$ and $\angle AMC = 30°$. Find the angle BAM.

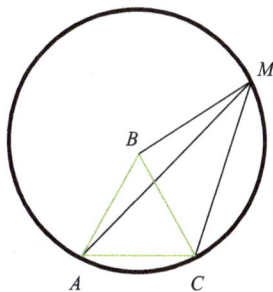

Solution. This problem is an excellent example of making use of an auxiliary circle. The clue to the shortest and the most elegant solution lies in the Inscribed Angle Theorem (an angle inscribed in a circle is half of the central angle that subtends the same arc on the circle). Someone can argue that there isn't a circle or any inscribed angles provided in the problem. This is true, but if you take a careful look at the diagram, then you should notice that both angles ABC and AMC are subtended by the same segment AC. Also, $\angle ABC = 2\angle AMC$ by the conditions of the problem. These two facts enable us to conclude that there is a circle with the center at B such that points A, C, and M lie on it. In this circle, angle ABC is the central angle, which is equal to $60°$, and AMC is an inscribed angle that is subtended by the same chord AC. Now we merely have to observe that being the radii of the same circle, $AB = BM$, and therefore, ABM is an isosceles triangle. Then $\angle BAM = \angle BMA = 20°$, and we are done.

Problem 3. In triangle ABC, angle B is $120°$. AA', BB', and CC' are the bisectors of angles A, B, and C respectively. Prove that $C'B'A'$ is a right angle.

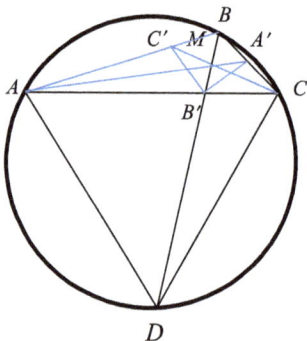

Proof. To solve the problem, obviously we should somehow use the properties of an angle bisector. Usually when we encounter a difficult problem like this, the first desire is to establish relationships between elements by using the theorem that the angle bisector of a triangle divides the opposite side into segments proportional in length to the adjacent sides. However, straightforward thinking here leads nowhere. After a couple of futile attempts students usually give up. The hint to this problem, as in many problems involving angle bisectors, is to make use of the definition of a bisector. Since each angle bisector divides an angle into equal parts, it would be a good idea to look for as many equal angles as you can. And here the circumcircle of the triangle might be a big help.

Remember, the inscribed angles subtended by the same chord or congruent chords are congruent to each other. The more inscribed angles we find, the clearer the path becomes to the goal. So, let us draw the auxiliary circumcircle of triangle ABC and then draw the bisector BB' of angle ABC. Denote D the point of intersection of the circle with the extension of BB'. By the Inscribed Angle Theorem, $\angle ACD = \angle ABD = 60°$ and $\angle CAD = \angle CBD = 60°$, because both pairs of angles are subtended by the same chord, AD and DC, respectively. This means that ADC is an equilateral triangle, hence

$$AC = AD = DC. \tag{1}$$

Also, we can see that $\angle AB'B = \angle DB'C$ as vertical angles and $\angle BDC = \angle BAC$ (the inscribed angles are subtended by a common chord BC), which means that $DB'C$ and $AB'B$ are similar triangles since they have two respectively equal angles. Therefore, $AB/DC = BB'/B'C$, and substituting the expression of DC from (1), we get

$$AB/AC = BB'/B'C. \tag{2}$$

Now it is time to recall the property mentioned earlier that an angle bisector divides the opposite side into segments proportional in length to the adjacent sides. Since AA' is a bisector of angle BAC then $AB/AC = BA'/A'C$. Inserting the value of AB/AC from (2) into the last equality yields

$$BB'/B'C = BA'/A'C. \tag{3}$$

Applying the statement converse to the property of a bisector just used, it will not be hard for readers to prove that $B'A'$ is the bisector of angle $BB'C$ in the triangle $BB'C$. Analogously, $B'C'$ is the bisector of angle $AB'B$ in the triangle $AB'B$.

Since $\angle AB'C = 180°$, then
$\angle C'B'A' = \angle C'B'B + \angle BB'A' = \frac{1}{2}(\angle AB'B + \angle BB'C) = 90°$,
which completes the proof.

Note that we have proved an even stronger statement than was required. The fact that $\angle C'B'A'$ is a right angle is an immediate consequence of the fact that $B'C'$ and $B'A'$ are angle bisectors of the triangles into which triangle ABC is partitioned by the bisector dropped from vertex B. This is a pretty interesting result by itself.

Problem 4. The base of a triangle equals 20; the medians dropped to the two other sides equal 18 and 24 respectively. Find the area of the triangle.

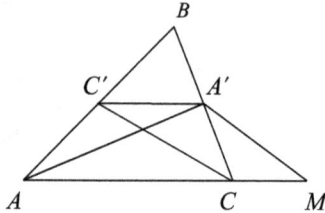

Solution. Assume that in triangle ABC, CC' and AA' are the medians, $AA' = 24$, $CC' = 18$, and $AC = 20$. We need to find the area of ABC.

First, note that $A'C'$ is the midline of triangle ABC, which means that $A'C' \parallel AC$ and $A'C' = \frac{1}{2}AC = 10$. These two facts give us a clue for the introduction of an auxiliary parallelogram. Let's construct $A'M \parallel CC'$. Then in the quadrilateral $C'A'MC$ the two pairs of opposite sides are parallel, which makes it a parallelogram. One may ask why did we draw $A'M$ parallel to CC'? Why do we need to introduce parallelogram $C'A'MC$?

Well, to answer these questions, let's analyze the problem. It is easy to notice that triangles ABC and $C'BA'$ are similar since they have equal angles. It means that the ratio of their areas equals to the squared ratio of their linear elements (for convenience, we can take the ratio $\frac{AC}{A'C'} = \frac{20}{10} = 2$). Now, denote the area of ABC by S and the area of $A'BC'$ by $\boldsymbol{S'}$. Then

$$\frac{S}{S'} = 2^2 = 4, \text{ or } S' = \frac{1}{4}S. \tag{*}$$

If we manage to express S' in terms of S in another suitable way, the problem will be solved. The easiest way to set up an equation is to find the difference $S - S'$, which is the area of the trapezoid $AC'A'C$. Now you can see how useful and helpful the introduction of an auxiliary element is. Since $C'A'MC$ is a parallelogram, its opposite sides are equal. Then $MC = A'C' = 10$ and $A'M = C'C = 18$. Obviously,

the area of triangle $AA'M$ equals the area of the trapezoid $AC'A'C$. I believe it is not going to be difficult for readers to prove this assertion by themselves. Just recall the formulas for the area of a trapezoid and the area of a triangle, expressed in terms of base and altitude dropped to its base. We know all the sides in triangle $AA'M$, so its area can be found by Heron's formula $S_{AA'M} = \sqrt{p(p-a)(p-b)(p-c)}$, where p is the semi-perimeter of the triangle and a, b, and c are its sides. Since $p = \frac{1}{2}(30 + 24 + 18) = 36$, then $p-a = 36-30 = 6$, $p - b = 36 - 24 = 12, p - c = 36 - 18 = 18$, and
$S_{AA'M} = \sqrt{36 \cdot 6 \cdot 12 \cdot 18} = 216$.

We see that $S - S' = S_{AC'A'C} = S_{AA'M} = 216$. By substituting S' from (*) in the last equality, we finally get $S - \frac{1}{4}S = 216$, from which $S = 288$, and we are done.

Problem 5. Point K lies on side AC of triangle ABC. AM is the median of ABC. Prove that the line passing through K and N, which lies on BC and is such that AN is parallel to KM, separates triangle ABC into two parts of equal area (Figure 1).

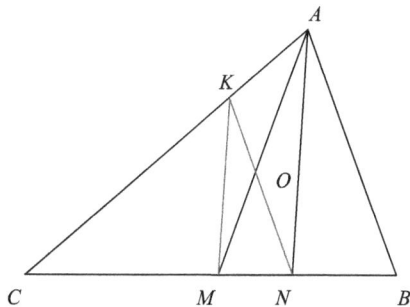

Figure 1

Proof. We must prove that the area of triangle KNC equals the area of the quadrilateral $KABN$.

First, notice that since a median of a triangle always divides it into two triangles of equal area, we have $S_{AMC} = S_{AMB} = \frac{1}{2}S_{ABC}$. Our plan then is to prove that the area of $KABN$ is equal to the area of AMB, which is half of the area of ABC. The quadrilateral $KABN$ consists of the quadrilateral $ABNO$ and the triangle KOA. On the other hand, the triangle AMB consists of the same quadrilateral

ABNO and the triangle *NOM*. Since *ABNO* is the common part of *KABN* and *AMB*, then, if we manage to prove that triangles *KOA* and *NOM* have equal areas, the problem will be solved.

Now it is the time to introduce an auxiliary element; of course, it's going to be a trapezoid. Indeed, both triangles *KOA* and *NOM* are parts of the quadrilateral *MKAN*, which is a trapezoid because *AN* is parallel to *KM*, and it would be natural to consider the properties of this figure. For convenience, let's draw a separate diagram of trapezoid *NMKA* (Figure 2).

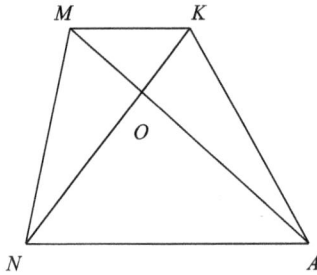

Figure 2

Earlier we proved (see Problem 4 in Chapter 6) that the areas of triangles *MON* and *KOA* are equal. To finish the solution of the problem, we recall that the area of the triangle *AMB* is half the area of *ABC*, and since the area of *KABN* equals the area of *AMB*, then the line *KN* divides the triangle into two parts of equal area, as we wished to prove.

Notice that in this problem, we utilized the auxiliary element differently than as in previous problems. In Problems 1 through 4, we transformed from the given elements to auxiliary elements and then made important observations by using the new figures. Next, we translated back the obtained results and made a conclusion in terms of the original problem. In Problem 5, we have built a logical chain step by step without using such transformations. If in Problems 1–4, one can say that auxiliary elements to some degree were introduced artificially, than in Problem 5, it was much easier to notice what should be used as an auxiliary element. We came across trapezoid *KANM* proceeding with our plan of proof, so I would say it was introduced more naturally.

Problem 6. Prove that three altitudes of a triangle meet at one point — the orthocenter of the triangle.

Proof.

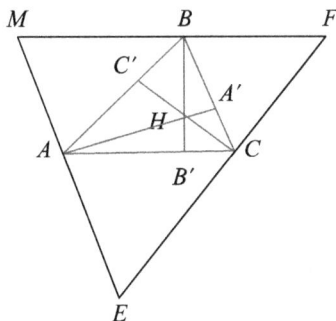

There are several proofs of this remarkable property, one of which was given earlier. Here we will explore another proof. In my opinion, one of the most elegant is *Carl Friedrich Gauss's* proof, using an auxiliary triangle. Before taking a closer look at this solution, let me point out and explain the main idea that underlies it.

In problem solving, it is very important to always remember that an element may play several roles and have many properties. We should apply such properties that are the most useful for particular purposes. For example, a median in an isosceles triangle dropped to its base might serve as an altitude and an angle bisector as well. Depending on the problem, we can apply the properties of a bisector and neglect the properties of an altitude or median, and vice versa.

In this problem, we are dealing with the altitudes of a triangle. The only fact that lies on the surface is that the segments BB', AA', and CC' are perpendicular to AC, BC, and AB respectively. However, it is not clear how we can use this in proving that the altitudes meet at one point. We should find some other properties of altitudes that might help us to succeed. The introduction of an auxiliary triangle makes that possible. If we draw parallel lines to the opposite sides of the triangle through the vertices A, B, and C, and designate the points of their intersections M, F, and E, then our altitudes will be necessarily perpendicular to the sides of the newly formed triangle MFE. And even more, they are perpendicular bisectors of the sides MF, FE, and ME. Indeed, in parallelograms $AMBC$ and $ACFB$, $MB = BF = AC$, which means that B is the midpoint of side MF (likewise, A is the midpoint of ME and C is the midpoint of EF). We know that three perpendicular bisectors of any triangle intersect at one point, the center of the circumcircle of that triangle. Hence, our proof is complete. Isn't it beautiful?

We did not prove that the three altitudes of the triangle meet at one point; instead by using the auxiliary triangle, we managed to prove that they are perpendicular bisectors in our auxiliary triangle. Therefore, they must be concurrent. The decision to introduce an auxiliary element in this problem naturally evolved from the property of a segment to be perpendicular to both parallel lines at the same time. It supports the advice given at the beginning of the chapter — look for the most essential properties of the given elements and then take advantage of them.

Problem 7. An equilateral triangle ABC is inscribed in a circle. A point M is randomly picked on arc BC. Prove that $MA = MB + MC$.

Proof. The clue to the solution of this problem lies in making use of the Inscribed Angle Theorem and the properties of equilateral triangles. Since both inscribed angles ACB and AMB are subtended by the same chord AB, then these angles are equal, which means that $\angle AMB = 60°$. And if we now pick D on the segment AM such that $MD = MB$, then triangle MBD will be equilateral.

Let's rotate triangle BCM about point B by $60°$. In this rotation, C is rotated into A (remember, ABC is an equilateral triangle, so all its angles are $60°$), M is rotated into D, and consequently, MC is rotated into DA, implying that $DA = MC$.

So, as you can see, the auxiliary triangle MBD helped us establish the fact that for any point M on the arc BC the required equality holds. Noticing that MA is the sum of MD and DA and substituting the respective values, we have $MA = MD + DA = MB + MC$.

By making use of an auxiliary triangle, we can obtain an interesting generalization of Problem 7:

For any randomly picked point M and an equilateral triangle ABC, there may exist a triangle with sides of length MA, MB, and MC, or one of these segments equals the sum of the other two (when point M belongs to circumcircle of triangle ABC).

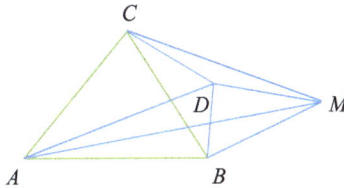

This statement is known as *Pompeiu's theorem*, after a Romanian mathematician Dimitrie D. Pompeiu (1873–1954).

Let's prove it.

Assume that M does not lie on the circumcircle of triangle ABC. Once again, by the rotation about B by $60°$ M will be rotated into D. But now D will not lie on line AM. Indeed, if for instance, M lies inside angle BAC but outside of the circumcircle of triangle BAC, then $\angle ADM = \angle ADB + 60° < 180°$, because $\angle ADB = \angle CMB < 120°$ (the rotation takes angle CMB into angle ADB, so they must be equal).

Note that $MD = MB$ and $DA = MC$, or in other words, the sides of triangle ADM have lengths MA, MB, and MC, which was to be proved.

To conclude the proof, it is necessary to consider all possible locations of M. I believe that will not be a hard task for readers to complete.

In Problem 7, the introduction of an auxiliary triangle gave us a purely geometrical solution — short, elegant, and vivid. It allowed us to not only simplify the solution but as well led us to a generalization, which covers all possible locations of point M.

The introduction of the auxiliary elements is one of the highest arts of the successful problem solver. The ability to recognize the essential and most important properties of the given elements is crucial for making efficient utilization of auxiliary elements. No one can guarantee that you will manage to find an advantageous auxiliary element on your first attempt. You should carefully analyze the problem and clarify for yourself the plan for the solution.

If you go back and review the problems covered throughout the book, you should see how the introduction of auxiliary elements simplifies some of the solutions. We did not emphasize that fact before this chapter, but I am sure you will recognize the application of the many auxiliary elements in problem solving in the previous chapters. To name a few: auxiliary segments led the way in the proof of Fagnano's Problem in Chapter 3 (instead of comparing the perimeters of the triangles, we compared the lengths of the auxiliary segment and the broken line); the auxiliary circle was utilized in the proof of Theorem 4 in Chapter 4; the auxiliary incircle and cyclic quadrilateral helped in proving Heron's formula in Chapter 5; and the auxiliary triangle with area equal to the area of the given trapezoid simplified the solution of Problem 5 in Chapter 7.

The method of auxiliary elements helps to stimulate imagination and creativity. It is a great example of how mathematicians use a variety of techniques to tackle problems.

Chapter 9

Constructions Siblings

Like crossword puzzles and jigsaw puzzles, constructions with a compass and straightedge are stimulating and engrossing. The most interesting and challenging construction problems are the so-called constructions with restrictions — problems in which the solution must be obtained exclusively by means of one tool, straightedge or compass. They encourage creative thinking, awaken the imagination, serve as motivating devices to supplement verbalization and lead to better understanding of the geometric principles involved.

"Like mother, like daughter" can analytically apply to mathematical principles where one mother problem can be the basic key to solving a chain, where all the problems following will be the construction siblings, and any one missing in the chain will make the solving of everything following very difficult. Problems later in the chain depend on those that precede them. Therefore, these problems form a logical chain reaction from one to another, requiring all previous steps and solutions.

Geometric figures are constructed by means of straightedge and compass. Since deductive reasoning is the basis for constructions, measuring instruments such as the ruler or protractor are not permitted. However, a ruler may be used as a straightedge by disregarding its markings.

In the following problems, a compass is not to be used; its lack makes the problems attractive and challenging.

Mother Problem. Given a trapezoid $ABCD$ ($AB \parallel CD$). P is the point of intersection of the non-parallel sides AD and BC, and O is the point of intersection of diagonals AC and BD. Prove that the straight line PO divides AB and DC into two equal parts.

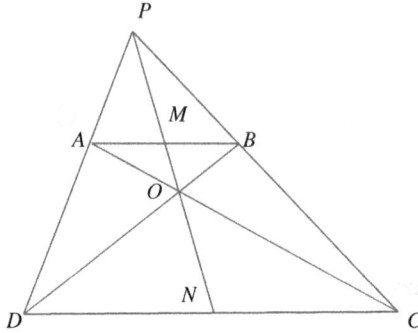

Proof.   Let's denote by M and N the points of intersection of lines AB and PO, and DC and PO respectively.

To solve the problem, we will use a succession of similar triangles (checking angles, it's not hard to see that the below pairs of triangles are similar):

From the similar triangles AMP and DNP, we have
$AM/DN = MP/NP$.

From the similar triangles MPB and NPC, we have
$MP/NP = BM/NC$.

From the similar triangles MAO and NCO, we have
$AM/NC = MO/NO$.

From the similar triangles MBO and NDO, we have
$MO/NO = BM/DN$.

Multiplying the four equalities and doing some cancellations, we are left with $AM^2 = BM^2$, from which $AM = BM$, implying that M is the midpoint of AB.

The proof that N is the midpoint of DC proceeds analogously.

We just proved the following powerful statement:

In a trapezoid, the midpoints of the bases are collinear with the point of intersection of the diagonals, and also with the point of intersection of the two non-parallel sides.

The reader may want to remember this important property. It comes in handy in solving many problems involving trapezoids.

Meanwhile, we proceed to apply it right away in Problems 2 and 3, which we call the daughters or offshoots of the Mother Problem.

Problem 2. Given a triangle ABC, BB' is a median and O is a random point on BB'. Lines CO and AO meet AB and BC at the points C' and A' respectively. Prove that $A'C' \parallel AC$ and BB' divides $A'C'$ into two equal parts.

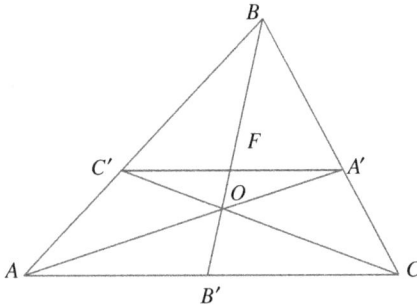

Proof. To prove the problem, we are going to use the equality $(AB'/B'C) \cdot (CA'/A'B) \cdot (BC'/AC') = 1$, which holds due to *Ceva's Theorem* (more on it later on in this chapter). It is given that $AB' = B'C$, hence, $\frac{CA'}{A'B} = \frac{AC'}{BC'}$, which yields $\frac{BC-A'B}{A'B} = \frac{AB-BC'}{BC'}$. Simplifying we have $\frac{BC}{A'B} - 1 = \frac{AB}{BC'} - 1$, and finally, $\frac{BC}{A'B} = \frac{AB}{BC'}$. We see that triangles ABC and $C'BA'$ are similar because they have a common angle B and corresponding sides in the same ratio. Therefore, $\angle BC'A' = \angle BAC$ and since these angles are alternate interior angles, $C'A' \parallel AC$. Because $AC'A'C$ is a trapezoid and F is the point of intersection of BB' and $A'C'$, by the Mother Problem, $C'F = FA'$, and we are done.

Problem 3. Given a trapezoid $ABCD$. Point F is the midpoint of side BC and point E is the midpoint of side AD. AF and BE meet at M, and AC and BD meet at O. Prove that $OM \parallel AD$.

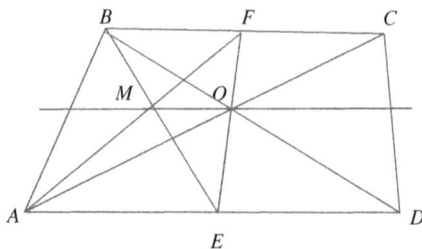

Proof. According to Problems 1 and 2, O lies on FE. For a proof, we have to observe that the pairs of similar triangles BMF and EMA, and BOF and DOE yield respectively $BM/ME = BF/AE$ and $BO/OD = BF/ED$.

Recalling that $AE = ED$, we obtain that $BM/ME = BO/OD$.

Angle B is a common angle of triangles MBO and EBD. The triangles also have corresponding sides in the same ratio. So, they are similar and therefore, the respective angles MOB and EDB are equal, and since they are alternate-interior angles, we get that $OM \parallel AD$, as requested.

The Mother Problem and its two offshoots are used as a foundation for solving the following chain of sibling problems. Remember that in every construction, we are to use the straightedge only.

Problem 4. Given a circle, two parallel chords, and a diameter that is not perpendicular to them, find the center of the circle.

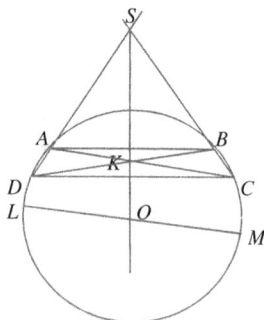

Solution. Assume LM is the given diameter and chord AB is parallel to chord DC. Extend lines DA and CB to their intersection at S, and draw the diagonals of the trapezoid $ABCD$, which intersect at K. The Mother Problem states that SK divides AB and DC into two equal parts. This means that SK contains the common median of the isosceles triangles ASB and DSC. We believe it would not be difficult for readers to prove on their own that ASB and DSC are isosceles triangles with equal sides SA and SB, and SD and SC, respectively. But in an isosceles triangle, the median drawn to the base coincides with the altitude dropped to that base. So, we obtain that $SK \perp AB$ and therefore, SK contains the diameter of the circle (perpendicular bisector of a chord of a circle passes through its center). The point O, where the lines SK and LM intersect, is the required center of the circle.

Problem 5. Given two parallel lines l and l'. Points A and B lie on the line l'.
(a) Divide the segment AB into two equal parts.
(b) Divide the segment AB into three equal parts.

Solution.

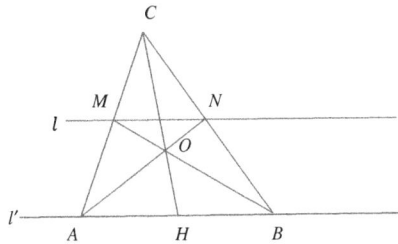

Figure 1

(a) In Figure 1, construct a random line through the point A meeting l at M. On the ray AM pick any point C. Draw the line BC. Let BC meet l at N. Then $AMNB$ is a trapezoid, the diagonals of which intersect at point O. According to the Mother Problem, the line CO must divide the segment AB into two equal parts, so $AH = HB$.

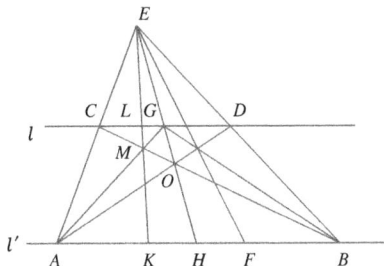

Figure 2

(b) In Figure 2, to divide the segment AB into three equal parts, we can use the construction explained above. Assume that AB has already been divided into two equal parts, $AH = BH$. $ACDB$ is a trapezoid. Let AC and BD meet at E, EH and l meet at G, AG and BC meet at M, and EM and AB meet at K. Considering the pairs of similar triangles CLE and AKE, CEG and AEH, CLM and BKM and remembering that $2AH = AB$, we obtain

$$CL/AK = EC/EA = CG/AH = 2CG/AB = 2CM/MB = 2CL/BK.$$

Therefore, $2AK = BK$, i.e., $3AK = AB$.

In the same way, we can find F where $2BF = AF$.

Thus, points K and F are the required points, $AK = KF = FB$.

It should be noted that in a similar manner, it is possible to divide a segment into n equal parts for any natural number n.

Problem 6. Given segment AB and its midpoint C. Point K does not lie on AB. Construct the line through K parallel to AB.

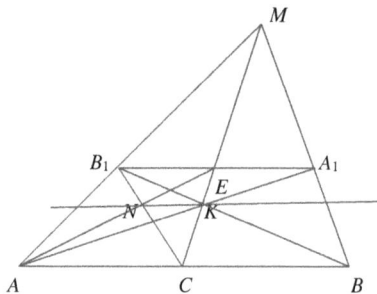

Solution. Draw CK. Pick any point M on CK. Draw AM and BM. Let BK and AM meet at B_1 and AK and MB meet at A_1. MC is the median in the triangle AMB and contains point K. From Problem 2, we know that $B_1A_1 \parallel AB$. Thus, AB_1A_1B is a trapezoid. Let MC and B_1A_1 meet at E and AE and B_1C meet at N. Using Problem 3, we see that $NK \parallel AB$, and therefore, NK is the required parallel.

Problem 7. Given two parallel lines m and n. Point P lies between them. Construct the line passing through P and parallel to m and n.

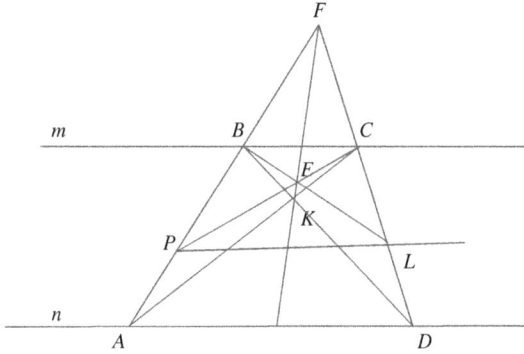

Solution. Draw any line through P that intersects lines m and n at points B and A, respectively. Pick any point F on that line ($F \neq A, F \neq B, F \neq P$) and draw another random line through F that intersects m and n at points C and D, respectively. Let AC and BD meet at K, FK and CP meet at E, and BE and FC meet at L. Using Problems 1 and 3, we conclude that PL is the required parallel.

Problem 8. Given a circle with the center O. Point M lies outside the circle. Construct the line through M parallel to a given line m.

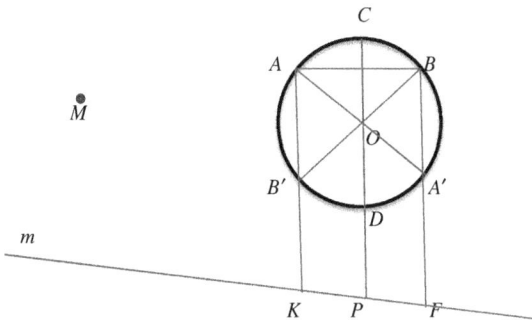

Solution. In figure above, draw a random chord AB and two diameters AA' and BB'. Let AB' and m meet at K, and BA' and m meet at F. Applying Problem 7, construct the diameter $CD \parallel AB'$. The extended line CD intersects m at P, where P is the midpoint of KF. It remains to draw the line passing at M parallel to FK, the midpoint of which P has already been found. This construction has been explained in Problem 6.

Problem 9. Construct the line through the point O parallel to the line m, if O is the center of the given circle.

Solution. When the line m does not intersect the circle or touches it, then all constructions are the same as in Problem 8. When m intersects the circle at two points, it's convenient to pick as a chord the segment between the points of intersection of line m and the circle. The problem then becomes Problem 8, and we can use the methods of construction outlined above.

Problem 10. Given a circle with center O, construct the bisector of the given angle ABC.

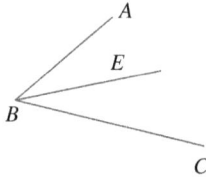

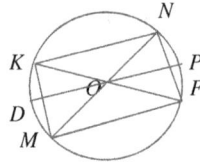

Solution. Draw diameters MN and KF, where $MN \parallel AB$, and $KF \parallel BC$. We can do this using Problem 9. $MKNF$ is a rectangle inscribed in the circle. Therefore, $NK \parallel MF$. Draw DP at O parallel to KN and MF (see Problem 7). OP is the bisector of the angle NOF. Recall that the sides of that angle are parallel to the corresponding sides of the angle ABC. The last step in our construction is to draw the line at B parallel to DP (again see Problem 7), $BE \parallel DP$. BE is the desired bisector of angle ABC.

The last construction problem of this chapter is done. At this point, it has to be emphasized that if the constructions are to be performed only by means of a marked ruler (no compass to be used), then it would be possible only to draw segments, whose lengths are expressed by rational operations and square roots of the sum of squares of already constructed segments.

Jacob Steiner

It is interesting to notice that every construction made by a straightedge and compass may be accomplished by means of a straightedge alone if there is a circle with its center provided as an additional condition of a problem. Problems 8–10 are the examples of such constructions. A prominent Swiss mathematician Jacob Steiner (1796–1863) managed to prove that all problems of the second order can be solved by means of the straightedge alone without the use of a compass as soon as a circle with its center is given as one of the problem's conditions. This assertion is known as *Poncelet–Steiner theorem.*

We have mentioned Jacob Steiner's name in Chapter 3 while studying one of his construction problems (Problem 1). He made tremendous contributions to the field of geometry and was considered by many to be one of the greatest geometers since ancient times. He deserves a particular note for his contributions related to projective geometry and modern synthetic geometry.

Another mathematician who will have to be mentioned is the Italian Giovanni Ceva (1647–1734), whose name is most often associated with the theorem we made use of in the solution of Problem 2.

In Ceva's honor, the line segments joining the vertices of a triangle to any points on the opposite sides are called *cevians*.

Giovanni Ceva

Even though Ceva's Theorem is not covered in high school, it provides a powerful and useful property of concurrent lines passing through the vertices of a triangle:

Given a triangle ABC, segments drawn from the vertices intersect the opposite sides at points M, L, and N. Lines AL, BM, and CN will be concurrent if and only if

$$\frac{AN}{BN} \cdot \frac{BL}{LC} \cdot \frac{CM}{MA} = 1. \tag{*}$$

Proof. Let's first prove the direct statement of the theorem: given that lines AL, MB, and CN meet at O, we need then to justify (*).

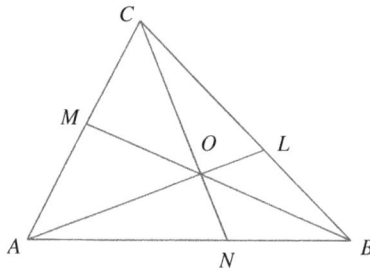

There are many proofs of this; we will use methods explained earlier. Applying the formula for the area of a triangle, $S = \frac{1}{2}ah$, it's easy to observe that for any two triangles with a common vertex and opposite sides that are bases on the same straight line, the ratio of their areas is the ratio of their bases. From the pairs of triangles ACN and BCN, AON and BON we get that

$$\frac{AN}{BN} = \frac{S_{ACN}}{S_{BCN}} = \frac{S_{AON}}{S_{BON}} = \frac{S_{ACN} - S_{AON}}{S_{BCN} - S_{BON}} = \frac{S_{AOC}}{S_{BOC}}. \tag{1}$$

Similarly,

$$\frac{BL}{LC} = \frac{S_{BOA}}{S_{COA}} \tag{2}$$

and

$$\frac{CM}{MA} = \frac{S_{COB}}{S_{AOB}}. \tag{3}$$

Multiplying (1), (2), and (3) we obtain the desired result:

$$\frac{AN}{BN} \cdot \frac{BL}{LC} \cdot \frac{CM}{MA} = \frac{S_{AOC}}{S_{BOC}} \cdot \frac{S_{BOA}}{S_{COA}} \cdot \frac{S_{COB}}{S_{AOB}} = 1.$$

The direct statement is justified.

Now we have to prove that if (*) is correct for the triangle ABC, then AL, BM, and CN intersect at the same point.

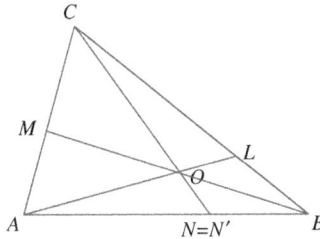

Denote the point of intersection of AL and BM by O. Draw CO so it intersects side AB at N'. Lines AL, BM, and CN' are concurrent at point O. Therefore, due to the just proved property $\frac{AN'}{BN'} \cdot \frac{BL}{LC} \cdot \frac{CM}{MA} = 1$ (**). Comparing (*) and (**) implies that $\frac{AN'}{BN'} = \frac{AN}{BN}$ or $\frac{AN'+N'B}{BN'} = \frac{AN+NB}{BN}$, which yields $\frac{AB}{BN'} = \frac{AB}{BN}$. Thus, $BN' = BN$ and points N' and N must coincide implying that the concurrency of lines AL, BM, and CN is proved.

The predecessor of Ceva's Theorem is Menelaus's Theorem named after a Greek mathematician and astronomer Menelaus of Alexandria (70–140 CE). Menelaus's Theorem concerns a line which intersects all three sides of a triangle. It proves very useful in solving problems involving the collinearity of three points.

The general case of the theorem through the concept of a Euclidean vector (a segment defined by its length and direction) states that

Given a triangle ABC with points D, E, and F on the side lines BC, CA, and AB, respectively, the points D, E, and F are collinear if and only if

$$\frac{\overrightarrow{AF}}{\overrightarrow{FB}} \cdot \frac{\overrightarrow{BD}}{\overrightarrow{DC}} \cdot \frac{\overrightarrow{CE}}{\overrightarrow{EA}} = -1.$$

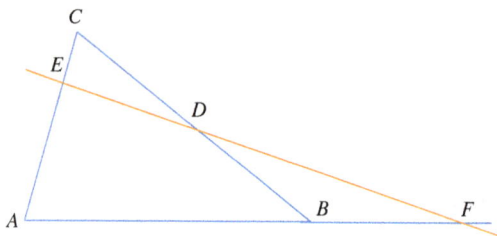

Interesting is that Giovanni Ceva, who rediscovered Menelaus's Theorem in 1678, proved this theorem and related Ceva's Theorem relying on the notion of center of mass of a system of three material points (see Chapter 1). Should it be a nice challenge for readers to recreate such a proof independently applying the principles outlined in Chapter 1?!

Ceva's theorem can be generalized to three and higher-dimensional simplexes (simplex is generalization of the notion of a triangle or tetrahedron to arbitrary dimensions):

The cevians are concurrent if and only if a mass distribution can be assigned to the vertices such that each cevian intersects the opposite facet at its center of mass. The intersection point of the cevians is the center of mass of the simplex.

The concept of designing a logical "chain reaction" discussed in this chapter may be extended to any type of problems, not just constructions. If you are able to find the mother who gives birth to multiple siblings, you may simplify the solutions, make them easy to follow, and utilize them in some other problems and projects.

As we discussed the problems–siblings relationships in this chapter, it is natural to desire to identify Ceva's Theorem as a Mother-Theorem for proving various cevians concurrencies. We have already examined the classical centers — points of concurrency of medians, altitudes, and angle bisectors in a triangle. The fact that all these segments in a triangle concur is truly amazing.

How about getting acquainted with another interesting and surprising sibling in this family of concurrencies? It will be used in the next chapter and is called the *Gergonne Point* problem:

The lines joining the vertices of a triangle to the tangent points of the inscribed circle with the opposite sides are concurrent at a point called the Gergonne Point.

The French mathematician Joseph Diaz Gergonne (1771–1859) was the first to establish this relationship. It is not a trivial problem to prove unless you rely on the Mother-Theorem, which enables you to come up with a pretty simple and elegant solution.

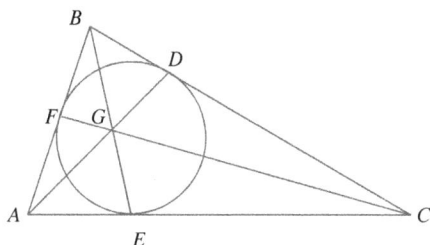

We merely need to observe three pairs of tangent segments of the same length:

$AF = AE$, $BD = BF$, and $CE = CD$. Multiplying these equalities will lead to $AF \cdot BD \cdot CE = AE \cdot BF \cdot CD$, which can be written as $\frac{AF}{BF} \cdot \frac{BD}{CD} \cdot \frac{CE}{AE} = 1$. By Ceva's Theorem, cevians AD, BE, and CF are concurrent, and we are done.

Having Ceva's Theorem in your arsenal, it would be a great exercise for you to investigate alternative proofs of the concurrency of the triangle's altitudes, medians, and angle bisectors. You should enjoy the brilliance of this property in giving relatively simple proofs of the orthocenter's, centroid's, and circumcenter's existence.

To conclude, there are many problems that lend themselves to solving by the methods described in this chapter. We restricted our attention to the several construction-siblings which have Problem 1 as their Mother. We also introduced Ceva's Theorem as another parent of the cevians concurrency problems. Readers may discover many new relatives of those families. The main idea of such discoveries is to train you to recognize which family a problem belongs to. Sibling problems help to develop logical thinking, give the opportunity to apply analytic and creative powers, and most importantly, encourage finding the links between problems in order to come up with the optimal solution.

Chapter 10

Session of One Interesting Construction Problem

How many solutions does a problem have? Why is it important to find different solutions to a single problem? Why aren't we satisfied sometimes with the solutions we've already found, and is it worthwhile to look for another one?

In finding multiple solutions to a problem, we can develop a deeper understanding of the subject matter. Searching for multiple solutions to a problem helps to develop problem-solving abilities and the flexibility that is often needed in a cooperative setting. The process can also stimulate the exploration of unfamiliar or little-known aspects of mathematics, as well as build a broader base of experience to draw on for more difficult problems and even generalizations. When you read a problem, you should begin to analyze the details and plan a strategy for the solution. You might assess the suitability of a technique by trying to visualize the result.

In this chapter, we pose one interesting construction problem and several approaches to its solution.

Problem. *Given an angle with an inaccessible vertex O, arbitrary points K and N on different sides of the angle, and a point M that is interior to the given angle. Construct the line passing through O and M.*

Before considering the solutions to the problem, I'd like to emphasize that the point O must not be used — it is inaccessible, and it is the lack of the ability to draw OM that makes the problem interesting. I believe that geometric construction problems, in which key parts of the figure are made inaccessible, are always challenging and captivating due to their non-standard character. What I find attractive in this problem is that one might have to solve it in real life, perhaps in constructing a building, in geological field work, or in land-surveying (geodesy industry).

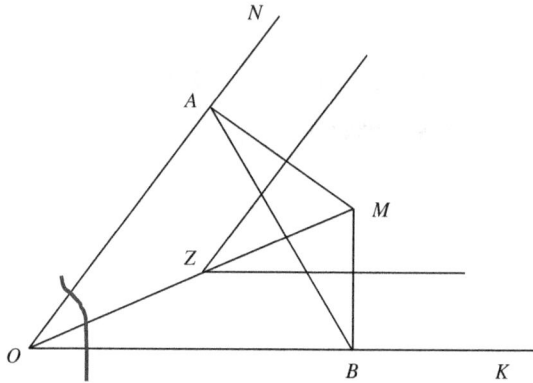

Solution 1. Draw the perpendicular to ON through M and denote the point of intersection by A. Draw the perpendicular to OK through M and denote the point of intersection by B. The right triangles OAM and OBM have a common hypotenuse OM; therefore, points A, M, B, and O lie on the circle whose center is at the midpoint of OM and whose radius is $\frac{1}{2}OM$. We now find the circumcenter of the triangle ABM, which coincides with the center of the circumscribed circle of triangles OAM and OBM. It is the point of intersection of the perpendicular bisectors of AM and BM — point Z. Thus MZ is the desired line.

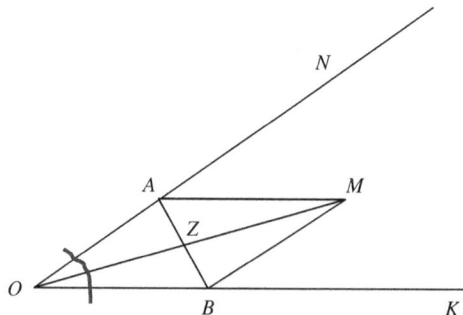

Solution 2. Draw $MA \parallel OK$ and $MB \parallel ON$.

Then $OAMB$ is a parallelogram by construction.

The diagonals of a parallelogram bisect each other. So if we find the midpoint of the diagonal AB (point Z), it must be also the midpoint of the diagonal OM.

Line MZ is the desired line.

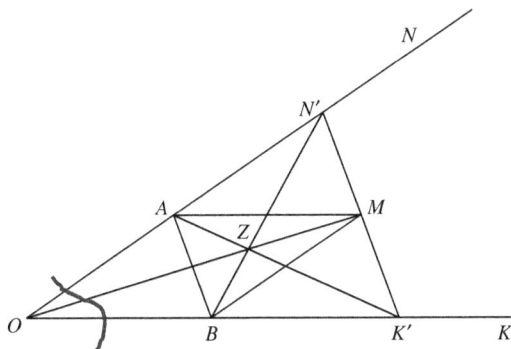

Solution 3. Draw $MA \parallel OK$ and $MB \parallel ON$. Draw $N'K'$ at M parallel to AB. Then we can show that M is the midpoint of $N'K'$. Indeed, by construction, $AN'MB$ is a parallelogram, hence $AB = N'M$; and $AMK'B$ is a parallelogram, hence $AB = MK'$. Thus $N'M = MK'$. Similarly, by choosing another pair of parallelograms, we can show that A is the midpoint of ON' and B is the midpoint of OK'. Thus $K'A$ and $N'B$ are medians of the triangle $ON'K'$. Their point of intersection Z is the centroid of the triangle, and its third median OM must also pass through Z. Therefore, MZ is the desired line.

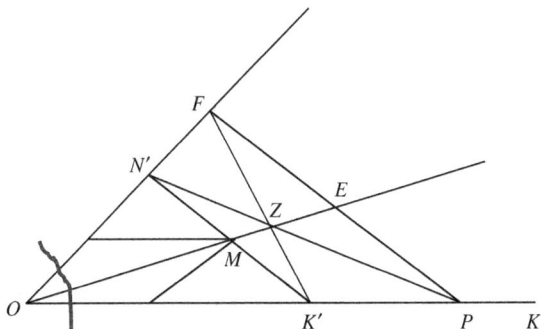

Solution 4. Using the method of Solution 3, we construct $N'K'$ through M such that M is its midpoint, and its endpoints are on the sides of the given angle. Draw an arbitrary segment FP parallel to $N'K'$ with its endpoints on the sides of the angle. Then $FN'K'P$ is the trapezoid, and point O is the point of intersection of its legs FN' and PK'. The Mother Problem from the chapter "Construction Siblings" says that in any trapezoid the midpoints of the bases, the point of the intersection of its diagonals, and the point of intersection of non-parallel sides are collinear. If Z is the point of intersection of $N'P$ and FK', then by the mentioned property from the Mother Problem, ZM will pass through point O. MZ is the desired line.

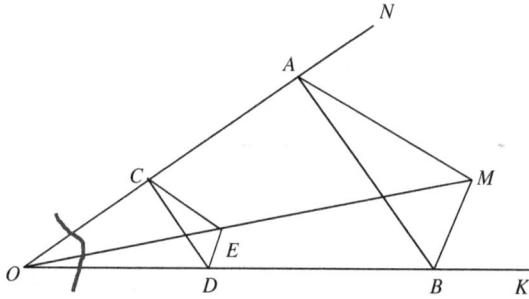

Solution 5. Construct an arbitrary triangle MAB with vertices A and B on the sides of the given angle ON and OK, respectively. Pick any point C on OA, and draw CD parallel to AB (where D lies on OB). Draw the line parallel to AM through C and also the line parallel to MB through D. Suppose the lines meet at point E. We will show that ME passes through O, solving our problem. Indeed, checking angles, we see that triangles COD and AOB are similar, hence $CD{:}AB = OC{:}OA = OD{:}OB$. Triangles CDE and ABM are similar as well, hence $CD{:}AB = DE{:}BM = CE{:}AM$. Thus,

$$CD{:}AB = OC{:}OA = OD{:}OB = DE{:}BM = CE{:}AM.$$

By construction, angles ODE and OBM are equal. Now, triangles ODE and OBM are similar because they have an angle in common, and the sides which include it are in proportion. This means that the respective angles DOE and BOM are equal as well. Since points O, D, and B are collinear, it follows that points O, E, and M must also be collinear, and ME is the required line.

Readers familiar with the notion of the homothetic transformations of Euclidean plane with respect to some fixed point must have recognized that triangle CED is the image of triangle AMB in the homothety with the center O and some ratio k, where $|k| < 1$. Clearly, we would be able to arrive at the same conclusion how to draw MO, relying on properties of homothetic transformations (under a homothety segments are mapped into parallel segments with a length which is $|k|$ times the original length).

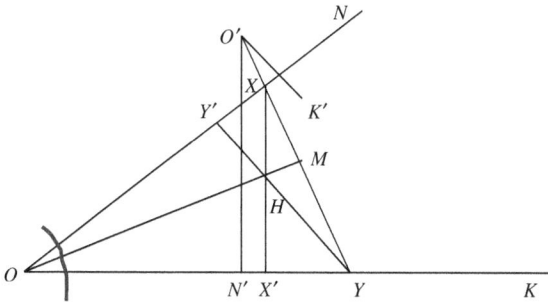

Solution 6. This solution uses rotation by $90°$ about M. Suppose such a rotation takes the angle KON into $K'O'N'$, where N' is on ray OK (note that this rotation can be performed without knowing where point O is, we rotate the accessible portions of the angle's sides). Then OM is perpendicular to MO', because the rotation preserves angles between corresponding lines, in this case $90°$. Suppose $O'M$ intersects ON at X, and OK at Y. Consider the triangle OXY, and draw its altitudes XX' and YY' which intersect at H. Since OM is perpendicular to XY (which is the same line as MO'), OM is the third altitude of the triangle. In other words, we obtain that H is the orthocenter of the triangle OXY, which implies that the line MH will pass through O. Thus, MH is the desired line.

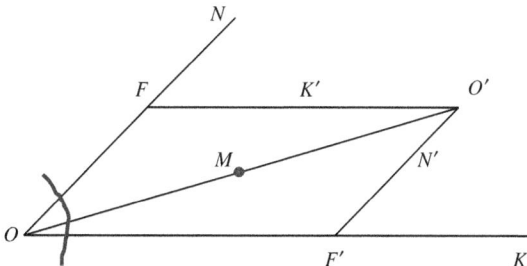

Solution 7. This time we rotate by 180° about the point M. The angle NOK (even if we cannot access its vertex) is taken into some angle $FO'F'$, where F is on ON and F' is on OK. (Again, we can construct the parallelogram without accessing O.) It follows that $OFO'F'$ is a parallelogram, and M is the midpoint of the diagonal OO'. Hence $O'M$ passes through O and solves our problem.

As you may have noticed, in every solution we introduced various auxiliary elements: circle, parallelogram, triangle, and trapezoid. We started with an auxiliary circle in Solution 1; let's then make a full circle in our journey and introduce another one in the final Solution 8.

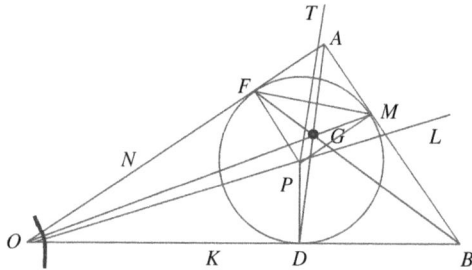

Solution 8. The goal in this solution will be to draw a circle inscribed in the angle NOK such that it passes through the given point M. Then we will construct the triangle with two sides on the rays ON and OK and the third side being tangent to the inscribed circle at M. In that triangle, we will be able to draw two cevians from the vertices to points of tangency with the inscribed circle. The point of their intersection will be Gergonne point of the triangle and, recalling that M is the third point of tangency, we conclude that MO must pass through this point, as was proved in the previous chapter.

Let's now follow our plan and do the constructions. First, select any point F on ON and draw FM. The center of the circle inscribed in the angle NOK and passing through M must be the point of intersection of the perpendicular bisector TP to FM and the angle bisector OL of $\angle NOK$ (you may go back to the solution of Problem 1 in Chapter 4 to see how to construct an angle bisector of an angle with an inaccessible vertex). As we get point P, the center of the circle, we draw the circle with radius $PM = PF$. The next step is to connect P and M and draw a line at M perpendicular to PM that intersects ON and OK at points A and B, respectively. As the result of all the work, we've done, the circle with center P is inscribed in the triangle AOB and points F and M are the points of its tangency with sides OA and BA, respectively. If we denote by D the point of tangency on OB, then by drawing AD and BF, we get their point of intersection G, which is the Gergonne point of the triangle AOB. The cevian connecting the inaccessible vertex O with M as the point of tangency of the incircle with AB must go through G. MG is the desired line.

As you can see, the methods and techniques suggested throughout the book have been broadly applied in approaching the above constructions. Properties of the orthocenter in Solution 6, properties of the perpendicular bisectors to sides of a triangle in Solution 1, properties of the centroid in Solution 3, properties of angle bisectors and methods outlined in "Constructions siblings" in Solutions 4 and 8, rotation and half-turn applications in Solutions 6 and 7.

It is worthwhile to compare the solutions offered here. In my opinion, the most elegant is Solution 1, but Solutions 2 and 7 are easier to follow and require fewer steps to construct; they also use more elementary ideas. Solutions 3, 4, 5, and especially 6 and 8 may seem more difficult. However, they are useful in strengthening thinking skills and creativity.

Note that Solution 4 was based on a Mother problem from the chapter "Constructions siblings". It is a good example of how the family relationships established in that chapter might lead to the discovery of interesting constructions while solving a problem not related to the siblings of that chapter. You never know when you may meet such a relative. The more you find, the better you are, and the more choices are available for you to come up with different solutions.

We learned from Solutions 3 and 6 how to construct a segment whose endpoints are on the sides of a given angle, passing through the given point inside the angle such that this point bisects the segment; and how to construct the line passing through a given point inside a given angle and perpendicular to the line connecting the vertex of the angle with the given point. I believe it would be a good exercise to find other solutions to these problems. Both problems were solved here with the restriction of using the vertex of the angle. You might get back to Chapter 7 and look at the solution of Problem 4 when the angle's vertex was available. During Solution 8, we learned how to construct the circle inscribed in the given angle and passing through a point lying inside of the angle. The Gergonne point, which you don't often come across or introduce as an auxiliary element, was a factor leading to the final step in the solution.

Finally, in discovering new solutions to the considered construction problem, you may come across some interesting and unexpected generalizations. A comparison of Solutions 1, 3, and 6 reveals the fact that in each of them, the desired line passes through one of the classic centers of a triangle — the circumcenter, centroid, and orthocenter. Is it always the case? Should it be applicable for any triangle? We dealt with different triangles, which all had in common just one angle with an inaccessible vertex. This interesting property has been already examined in Chapter 2, when we discussed the great discovery of the "master of all mathematicians", Leonhard Euler. It was proved that the circumcenter, centroid, and orthocenter of a triangle are collinear and the centroid divides the distance from the orthocenter to the circumcenter in the ratio 2:1. How about using our multiple solutions and arrive at the existence of the Euler Line and its properties from a different perspective?

The most difficult part of the problem-solving process is finding a route to link given facts and conditions together, and to identify how to use these facts in a logical chain to get to the desired outcome. Sometimes, the facts utilized in the solution of one problem serve as a flashlight that guides us in observing the conditions that can lead to new and amazing discoveries. Indeed, the constructions in Solutions 1 and 3 help us to notice that the perpendicular bisectors of the original triangle must contain the altitudes of its medial triangle. That simple fact leads to another interesting observation that the circumcenter of the original triangle, being the point of intersection of its perpendicular bisectors, coincides with the orthocenter of the medial triangle, which would immediately get us to another very interesting conclusion ... But let's be patient and do it step by step.

Consider triangle ABC with its orthocenter H, centroid O, and circumcenter P. Points F, K, and E are the midpoints of the sides AB, AC, and BC respectively. $BD \perp AC$ and $LK \perp AC$. The goal is to show that points H, O, and P are collinear.

In figure below, the perpendicular bisectors of the original triangle ABC must contain the altitudes of its medial triangle FEK. Therefore, the circumcenter P of the original triangle coincides with the orthocenter of the medial triangle. Obviously, triangles ABC and

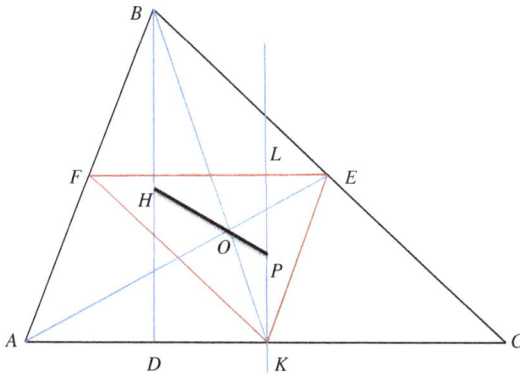

EKF are similar with the ratio 2:1 (see Chapter 1). BH and KP are corresponding segments in those triangles (segments connecting the corresponding vertices of triangles to the orthocenter of each triangle). Then $BH = 2KP$. Notice also that $BO = 2OK$ because the centroid divides medians in the ratio 2:1. Finally, $\angle HBO = \angle PKO$ because they are interior angles formed by the intersection of BK with two parallel lines BH and KL (BH lies on the altitude BD and KL is the perpendicular bisector to side AC). Thus, triangles BOH and KOP are similar, which means that the respective angles HOB and POK are congruent making them vertical angles, and we conclude that points P, O, and H must be collinear. The centroid lies between the circumcenter and orthocenter and is twice as far from the orthocenter as it is from the circumcenter (this follows from the similarity of triangles BOH and KOP with the ratio 2, i.e., $OH = 2OP$, as the corresponding sides in these similar triangles).

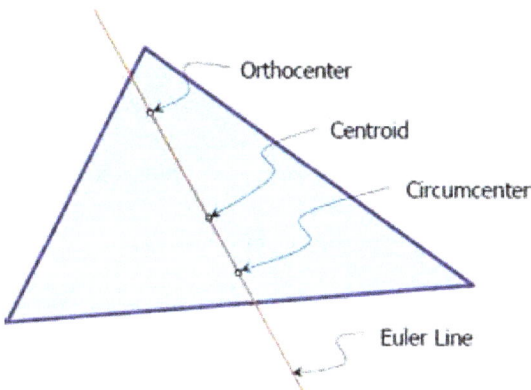

As we did not intend even to mention the Euler Line when exploring various approaches to the construction problem, it is a fascinating coincidence that Leonhard Euler did not mean to prove that the three classic centers of a triangle form one straight line either. He came across his famous discovery incidentally while solving a different problem:

Given four points: circumcenter, centroid, orthocenter, and incenter of a triangle, reconstruct the triangle.

Leonhard Euler

An interesting analogy between Euler's line discovery and Columbus's discovery of America was made by Professor Ed Sandifer in his work *The Euler Line* (*MAA Online, How Euler Did It.* January 2009). He said that both made their discoveries while looking for something else. "Columbus was trying to find China. Euler was trying to find a way to reconstruct a triangle, given the locations of some of its various centers. Neither named his discovery. Columbus never called it 'America' and Euler never called it 'the Euler Line'. Both misunderstood the importance of their discoveries".

In conclusion, I want to emphasize that the search for multiple solutions should whet readers' appetites and motivate discussions. It helps to solidify their grasp of mathematical ideas and offers further explorations on specific topics. It serves as an excellent exercise, which challenges them, fostering problem-solving abilities. Who knows, maybe one day you would be able to have your own discovery of some new geometric line or the new world.

Chapter 11

Alternative Proofs of the Pythagorean Theorem

Did not we have fun exhibiting multiple solutions to the interesting problem in the previous chapter? Speaking about alternative proofs of a problem, one of the most famous theorems in Euclidean geometry, the Pythagorean Theorem, naturally comes to mind.

It has more than 400 different proofs, perhaps more than any other theorem.

We will concentrate here on a few proofs reinforcing various strategies and techniques covered in previous chapters. Also, we will cover in more detail one very powerful technique for solving geometric problems briefly mentioned in Chapter 2, the so-called method of areas reasoning.

The Pythagorean Theorem:

Prove that in a right triangle the sum of the squares of the measures of the legs equals the square of the measure of the hypotenuse.

Proof 1. Applying the Theorem of Ratios of the Areas of Similar Polygons.

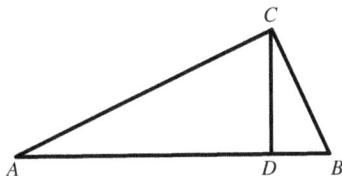

In right triangle ACB ($\angle ACB = 90°$), draw $CD \perp AB$. CD divides ACB into two right triangles similar to each other and to the big one.

Right triangles ADC and ACB, and BDC and BCA are similar because they share acute angle A and acute angle B respectively. Let's denote the area of the triangle ABC by S, the area of the triangle ADC by S', the area of the triangle BDC by S'', and the lengths of the sides $BC = a$, $AC = b$ and $AB = c$. Applying the Theorem of Ratios of the Areas of Similar Polygons to the pairs of similar triangles gives

$$\frac{S'}{S} = \frac{b^2}{c^2} \quad \text{and} \quad \frac{S''}{S} = \frac{a^2}{c^2}.$$

Adding the equalities leads to $\frac{S'+S''}{S} = \frac{b^2+a^2}{c^2}$.

Because $S' + S'' = S$, $\frac{b^2+a^2}{c^2} = 1$, from which $a^2 + b^2 = c^2$, and we are done.

Proof 2. Utilizing "basic" similar triangles.

In many geometric problems, identifying basic congruent or similar triangles is the very first step helping to connect given and known elements with unknown.

Referring to the diagram from the previous proof, we will consider similar right triangles

$$ACB \sim ADC \sim CDB.$$

Using the same notations for the lengths of the sides, $BC = a$, $AC = b$, and $AB = c$ and denoting $AD = h_a$ and $BD = h_b$ we establish the following relationships:

$$\frac{AC}{AB} = \frac{AD}{AC} \tag{1}$$

$$\frac{BC}{DB} = \frac{AB}{BC} \tag{2}$$

Substituting the respective values into (1) and (2) we rewrite them as $b^2 = c \cdot h_a$ and $a^2 = c \cdot h_b$. Adding these two equalities and noticing that $h_a + h_b = c$ gives $b^2 + a^2 = c(h_a + h_b) = c^2$. We arrive at the desired result $a^2 + b^2 = c^2$.

Proof 3. Utilizing Tangent-Secant Theorem (referring to similar triangles one more time).

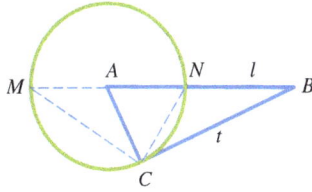

The Tangent-Secant Theorem asserts that given a secant l intersecting the circle at points M and N and tangent t touching the circle at point C and given that l and t intersect at point B, the following equality holds: $BC^2 = BM \cdot BN$. This can be easily proved utilizing similar triangles BCN and BMC (we leave this proof to the reader or you can find a proof in a standard geometry textbook).

Consider right triangle ACB ($\angle ACB = 90°$), and denote the lengths of its sides $BC = a, AC = b$, and $AB = c$. Draw a circle centered at A with radius $r = b$ such that straight line BA (its extension) intersects the circle at points M and N, and BC is the tangent touching this circle at C. Observe that in our nominations, $BM = c + b$ and $BN = c - b$.

Now, applying the Tangent-Secant Theorem yields $a^2 = (c+b)(c-b)$, from which we derive the desired equality $a^2 + b^2 = c^2$.

In the next two proofs, we will demonstrate the application of the "method of areas reasoning". Method of areas reasoning might be very efficient and useful in those cases when comparing related areas allows arriving at relationships of linear elements in question. This method requires creativity and "outside of the box" thinking because it often concerns a comparison of related areas in problems that originally have nothing to do with "direct" calculations of areas of the figures involved.

Proof 4. Method of area reasoning (trapezoid).

Here, we will demonstrate an interesting approach involving the areas calculations of right triangles comprising a right trapezoid. It is one of many so-called rearrangement proofs of the theorem.

This elegant proof belongs to James Abram Garfield (1831–1881) the 20th president of the United States of America. It was published in the *New England Journal of Education* in 1876. What is quite impressive about this proof is that it was discovered by a prominent Politian who was not a mathematician!

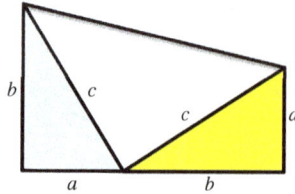

The trapezoid above is partitioned into two congruent blue and yellow right triangles and white isosceles right triangle which has the lengths of its legs equal to the length of the hypotenuse of blue and yellow triangles. The lengths of the sides have been labeled a, b, and c.

The area of a right triangle is calculated as half the product of its legs. Thus, the areas of the blue, yellow, and white triangles are respectively $S_1 = S_2 = \frac{1}{2}ab$ and $S_3 = \frac{1}{2}c^2$.

Observing that the altitude of our trapezoid has length $h = a + b$, we can express its area as
$S = \frac{1}{2}(a+b)h = \frac{1}{2}(a+b)(a+b) = \frac{1}{2}(a+b)^2$.

Since the area of the trapezoid equals the sum of the areas of the three right triangles into which it was partitioned, we get
$\frac{1}{2}ab + \frac{1}{2}ab + \frac{1}{2}c^2 = \frac{1}{2}(a+b)^2$.

Doing simple algebraic manipulations, we have
$ab + \frac{1}{2}c^2 = \frac{1}{2}a^2 + ab + \frac{1}{2}b^2$. Cancelling out like terms and multiplying both sides of this equality by 2 we arrive at the desired relationship $a^2 + b^2 = c^2$.

Proof 5. Method of area reasoning (square).

Consider a right triangle with the legs a and b and hypotenuse c. We arrange four such congruent triangles to form two squares, the big one with the side c and the small one inside it, with the side $b-a$.

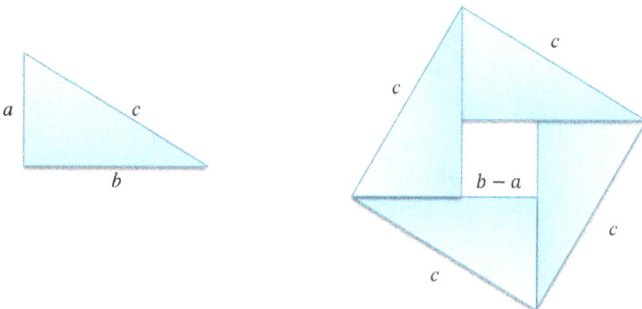

Applying simple algebraic techniques, we can easily get the result comparing the areas of the two squares and four right triangles on the right diagram above. The area of the big square is

$$S = c^2. \tag{1}$$

On the other hand, the same area can be calculated as the sum of the areas of four congruent right triangles and the small square inside the big one. The area of each of our four triangles equals half the product of its legs, $S_\triangle = \frac{1}{2}ab$. The area of the inside small square equals $(b - a)^2$.

It follows that the area of the big square can be calculated as

$$S = 4 \cdot \frac{1}{2}ab + (b - a)^2 = 2ab + b^2 + a^2 - 2ab = b^2 + a^2. \tag{2}$$

Equating (1) and (2) results in

$$a^2 + b^2 = c^2.$$

The next two proofs are not as elegant and easy to follow as the proofs above, but both present nice purely geometric proofs concerning the properties of triangles, rectangles, trapezoids, and squares.

These proofs are related to the original statement of the theorem how it is known since ancient Greek times:

The area of the square built on the hypotenuse is equal to the sum of the areas of the squares built on the two legs.

Proof 6. Applying rotations and auxiliary elements.

Draw three squares $ACED$, $BCFK$, and $ABNM$ on the sides of the right triangle ABC ($\angle C = 90°$). The goal will be to prove that the area of the square $ABNM$ equals the sum of the areas of the squares $ACED$ and $BCFK$.

First, counting the angles by vertex C ($\angle DCA + \angle ACB + \angle BCK = 45° + 90° + 45° = 180°$) we easily establish the collinearity of points D, C, and K.

Next, draw the diagonals AN and BM of $ABNM$ and denote O the point of their intersection.

For the proof, we will apply half-turn with respect to the center of rotation O and rotate $\triangle ABC$ into $\triangle NML$. Being the images in this rotation, these triangles are congruent, which immediately implies that two auxiliary quadrilaterals $ACLM$ and $ADKB$ are congruent as well (just compare the lengths of the respective sides and measures of the respective angles). Observe that $ADKB$ consists of right triangles DAC, ACB, and CBK. The areas of DAC and CBK represent one half of the areas of the squares $ACED$ and $BCFK$ respectively.

Therefore, finding the area of $ADKB$, we have

$$S_{ADKB} = \frac{1}{2}S_{ACED} + S_{ACB} + \frac{1}{2}S_{BCFK}. \tag{1}$$

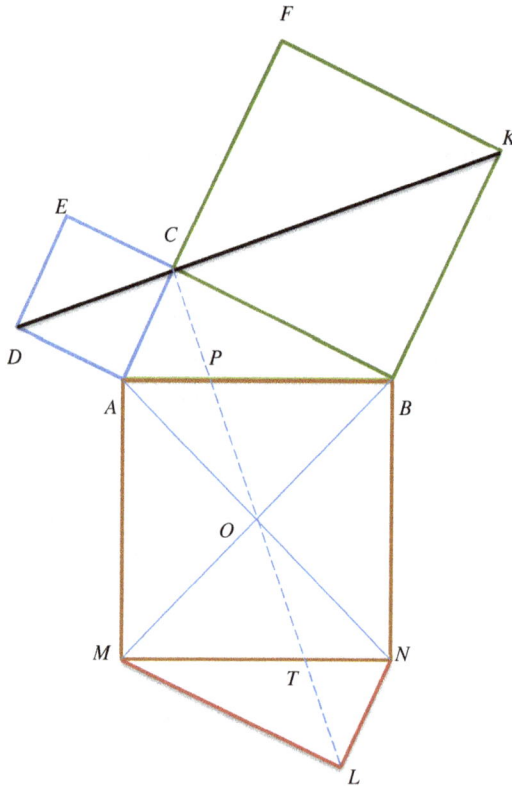

Now, notice that in our rotation the points of intersection of CL with AB and MN, P and T respectively, are the images of each other. It follows that there are two pairs of congruent triangles $\triangle ACP \cong \triangle NLT$ and $\triangle BPC \cong \triangle MTL$. This implies that $ACLM$ consisting of ACP, $APTM$, and MTL, in fact, is partitioned into the figures comprising triangle ACB (ACP and MTL) and half of the square $ABNM$ (the proof of the fact that $APTM$ is congruent to

$NTPB$ is left to the reader). Therefore, the area of $ACLM$ can be calculated as $S_{ACLM} = \frac{1}{2}S_{ABNM} + S_{ACB}$.

Finally, recalling that we proved earlier that $ACLM$ and $ADKB$ are congruent, and comparing the last equality with (1) leads to

$$\frac{1}{2}S_{ACED} + S_{ACB} + \frac{1}{2}S_{BCFK} = \frac{1}{2}S_{ABNM} + S_{ACB}.$$

Cancelling out S_{ACB} and multiplying by 2, we arrive at $S_{ABNM} = S_{ACED} + S_{BCFK}$, which is the desired result.

One final remark — the two auxiliary quadrilaterals allowed us to connect the areas of the squares in question without doing any algebraic calculations. All we had to do was to recognize and "gather" in a specific way the pieces into which both quadrilaterals were partitioned on our diagram. Comparing their areas, we arrived at the desired outcome.

Proof 7. Here we will discuss the classic Euclid's proof of the theorem.

The goal will be to prove that the area of the blue square equals the sum of the areas of the green and orange squares depicted on the diagram below.

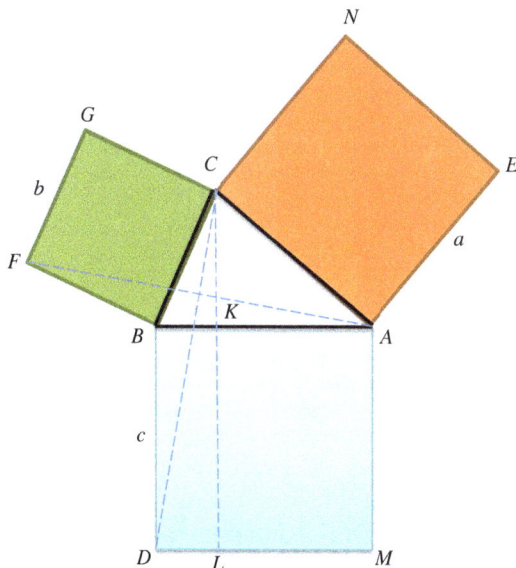

Draw $CK \perp AB$ and extend it till intersection at L with the side of the blue square DM. By doing this, we partitioned the blue square $AMDB$ into two rectangles $BKLD$ and $AKLM$. We will prove that the area of each of these two rectangles equals respectively to the area of the green square $BFGC$ and the area of the orange square $ACNE$. This will enable us to get to the desired result by just adding the areas of two rectangles and the respective squares.

First, let's connect A and F and C and D, and prove the congruency of the newly formed triangles FBA and CBD.

Notice that $BA = BD$ and $BF = BC$ as the sides of the blue and green squares respectively. Also,
$\angle ABF = \angle ABC + \angle FBC = \angle ABC + 90°$ and
$\angle CBD = \angle ABC + \angle ABD = \angle ABC + 90°$. Therefore,
$\angle ABF = \angle CBD$, and we conclude that triangles FBA and CBD are congruent by SAS property.

To simplify matters, we will show below the excerpt from the above diagram, depicting only triangle CBK and rectangle $BKLD$.

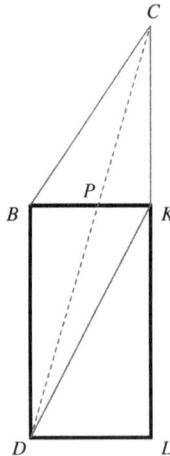

K lies on CL and $CL \parallel BD$. Therefore, $CKDB$ is a trapezoid. Denote P the point of intersection of the diagonals CD and BK. Triangles BPC and KPD have equal areas (see Problem 4 in Chapter 6), $S_{BPC} = S_{KPD}$.

We have $S_{CBD} = S_{BPC} + S_{BPD}$, and substituting the area of KPD for the area of BPC, leads to

$$S_{CBD} = S_{KPD} + S_{BPD} = S_{DBK} = \frac{1}{2} S_{BKLD}.$$

Getting back to the original diagram, in a similar fashion, we can prove that the area of the triangle FBA equals half the area of the square $BFGC$. Recalling earlier proven fact that triangles FBA and CBD are congruent and therefore, they have equals areas, we conclude that rectangle $BKLD$ and square $BFGC$ are equal in area. Similarly, we can get that $AMLK$ and $ACNE$ are equal in area as well.

Finally, substituting the respective equal areas in the below equality, we obtain

$$S_{AMDB} = S_{AMLK} + S_{BKLD} = S_{ACNE} + S_{BFGC}.$$

The theorem is proved.

It is important to mention that the Pythagorean Theorem is a two-way theorem. The converse of the Pythagorean Theorem is also true:

Given a triangle with sides of length a, b, and c, if $a^2 + b^2 = c^2$, then the angle between sides a and b is a right angle.

In other words, for any three positive real numbers a, b, and c such that $a^2 + b^2 = c^2$, there exists a right triangle with sides a, b, and c, where c represents the length of a hypotenuse and a and b are the lengths of the legs.

The proof of the converse statement entails using the Law of cosines.

Assume that in a triangle, the lengths of the sides satisfy the equality $a^2 + b^2 = c^2$.

On the other hand, by the Law of cosines,
$a^2 + b^2 - 2ab \cdot \cos \gamma = c^2$, where γ is the angle between sides a and b.

Comparing these equalities, we conclude that for the considered triangle, $2ab \cdot \cos \gamma = 0$. Since $a \neq 0$ and $b \neq 0$ it follows that $\cos \gamma = 0$, i.e., the angle between a and b has to be a right angle, $\gamma = 90°$, which is what was to be proved.

In conclusion of our discussion, let's look at some of the theorem's generalizations.

One such nice generalization was presented by Problem 10 in Chapter 6:

In a right triangle, the altitude dropped to the hypotenuse divides it into two triangles such that for any corresponding linear elements of the two triangles and the original triangle,

$$l^2 = l_1^2 + l_2^2.$$

Another interesting generalization can be derived considering the areas of any similar concave polygons on the three sides of a right triangle:

The sum of the areas of the similar polygons built on the two legs is equal to the area of the similar polygon built on the hypotenuse.

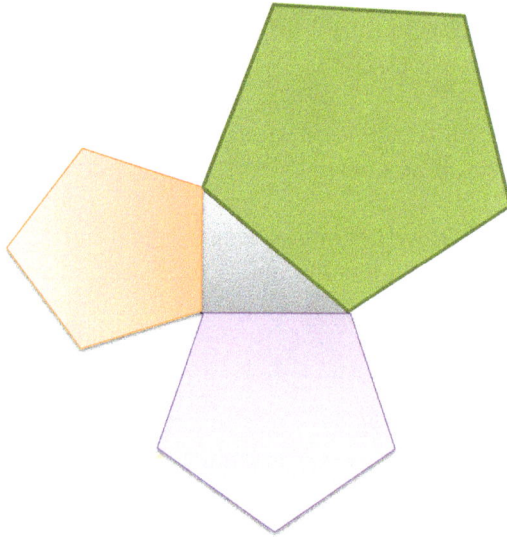

The above assertion that the area of the green polygon is equal to the sum of the areas of orange and purple polygons (pentagons in the figure above) can easily be proved applying the Theorem of Ratios of the Areas of Similar Polygons, and it should be a good exercise to try on your own.

The theorem can be extended to any similar figures that have curved boundaries (but still with part of a figure's boundary being the side of the original triangle).

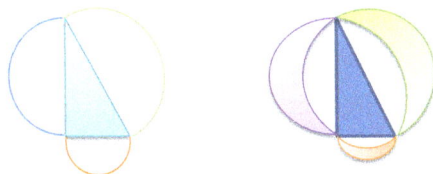

These properties should be very useful in tackling many Olympiad type problems concerning combinations of similar figures constructed on the sides of a right triangle.

Speaking about numerous applications of the Pythagorean Theorem, we ought to mention that it is a starting place and the basis for all of trigonometry.

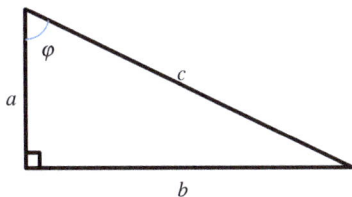

In a right triangle with legs a and b and hypotenuse c, trigonometric functions sine and cosine of the angle φ between leg a and the hypotenuse c are defined as $\sin \varphi = \frac{b}{c}$ and $\cos \varphi = \frac{a}{c}$. From this definition, we immediately arrive at the fundamental trigonometric Pythagorean Identity $\sin^2 \varphi + \cos^2 \varphi = 1$.

The Pythagorean Theorem gave the name to the famous Pythagorean triples consisting of three positive integers a, b, and c, such that $a^2 + b^2 = c^2$. Looking for integer triples satisfying this equation, we, in fact, are solving a *Diophantine equation* (an equation with two or more variables whose values are restricted to integers; Diophantus of Alexandria was the first to study such equations in the 3rd century AD), one of the oldest algebraic problems known. This topic alone could be a subject matter of a new book.

The Pythagorean Theorem is the basis for many fields outside of math, including physics, engineering, geology, architecture, navigation, surveying, and computer science to name a few.

Alternative proofs of the Pythagorean Theorem provide a seemingly endless array of geometric topics to talk about. One excellent source is Elisha Scott Loomis's book *The Pythagorean Proposition* (Reston, VA: NCTM, 1968) which presents 370 proofs.

Searching for multiple alternative proofs of the one of the most famous theorems in all of mathematics exposes you to many fascinating topics in Euclidean geometry. It should also greatly benefit sharpening your problem-solving skills.

How delightful and inspiring would it be to find your own proof of the Pythagorean Theorem?!

Chapter 12

Morley's Theorem

It is one of the most astonishing and totally unexpected theorems in mathematics and, jewel that it is, for sheer beauty it has few rivals.

Cletus O. Oakley and Justine C. Baker
"The Morley Trisector Theorem"

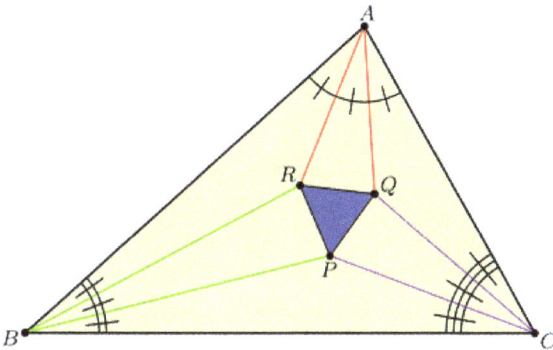

This is going to be the final chapter in the *Geometrical Kaleidoscope.* What does the word *kaleidoscope* mean? A kaleidoscope generates a beautiful and mesmerizing display of images created by the reflection of light. In our case, *Geometrical Kaleidoscope* was supposed to produce a rainbow of challenging and fascinating facts and problems from plane geometry and demonstrate how instrumental it is to use various methods and techniques in order to find the simplest and the most intuitive arguments to solve a variety of problems. In the end, we saved a real mathematical jewel — one of the most astonishing and intriguing problems in plane geometry, the theorem known as

Morley's Miracle or *Morley's Theorem*. Its discovery belongs to the professor of mathematics at Haverford College, Frank Morley (1860–1937).

The statement of the theorem is amazingly simple:

The three points of intersection of the adjacent angle trisectors of any triangle form an equilateral triangle.

In the triangle ABC, the trisectors (the lines dividing the angles into three equal parts) AR, AQ, CQ, CP, BP, and BR meet at the three points R, Q, and P, which will be the vertices of an equilateral triangle no matter what the angle measures of the original triangle are.

What is truly amazing is that this property was overlooked for so many years and its discovery was not made until 1899. The 13 books of *Euclid's Elements*, the most successful and influential textbook in plane geometry date back to 300 BC. Isn't it amazing that such a remarkable property was missed and not discussed up until the end of the 19th century? Euclid apparently did not ask how the trisectors of a triangle meet. The ancient Greeks deeply studied and developed various constructions by means of straightedge and compass. The angle trisection problem is impossible to solve with only those two classic construction tools. Therefore, it is probable that not much attention was devoted to trisectors in any construction, and general results concerning them were deemed uninteresting. There is no doubt that the Greeks could have proven Morley's Theorem with their knowledge and the numerous mathematical methods they had in the Euclidean times. They just hadn't noticed it. Actually, nobody saw it for more than 2000 years. Morley's astonishing discovery is evidence of not just his mathematical genius, but also his ability to see the world around him from a different perspective.

As simple as it looks, Morley's Theorem is not an easy theorem to prove; as a matter of fact, it is one of the toughest.

Over the years, there have been developed many proofs of the theorem: trigonometric proofs (the congruence of angles in Morley's triangle is achieved by way of trigonometric manipulations and calculations), backward proofs (where the proof starts with an equilateral triangle and shows that a triangle may be built around it that will be similar to selected triangle), algebraic proofs, and synthetic proofs. In 1913, Taylor and Marr presented the first complete proof, a pure geometric proof, which we are going to discuss below. It was not selected just because it was the first. It is an excellent demonstration of many methods and techniques applied throughout the book: properties of angle bisectors, reflections, cyclic quadrilaterals, inscribed angles in a circle subtended by the same chord, and properties of right triangles.

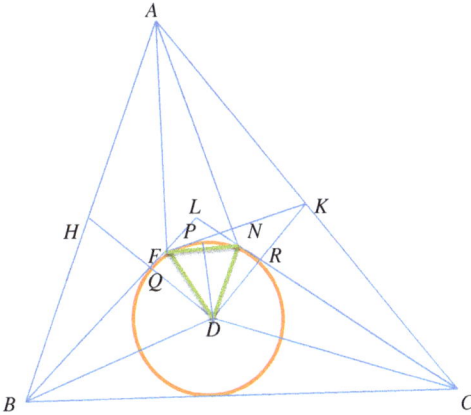

Figure 1

Let's consider the triangle ABC and denote by α, β, and γ the angles representing $\frac{1}{3}$ of the measure of each angle $\frac{A}{3}$, $\frac{B}{3}$, and $\frac{C}{3}$ respectively. Then

$$\alpha + \beta + \gamma = 60°. \tag{*}$$

Draw the trisectors of angles B and C. Points L and D are the points where they meet. Since D is the point of intersection of the angle bisectors in triangle BLC, it has to be the center of its inscribed circle. The next step will be to draw the images H and K of D by reflection in BL and CL respectively. If Q and R are the points of intersection of DH and DK with BL and CL, then they must be the points of tangency of the incircle with center D, and $DQ = DR$ as the radii of that circle. Then by the properties of reflections, $DQ = QH = DR = RK$ and we may conclude that $DH = DK$.

Now, draw the circle's tangent KP (P is the point of tangency) to its intersection with BL at point F.

Since P is the point of tangency of line KP with the incircle of the triangle BLC, then $DP \perp PK$. Furthermore, $PD = DR = r = \frac{1}{2}DK$, which implies that in the right triangle DPK, $\sin \angle K = \frac{1}{2}$ and therefore, $\angle K = 30°$; yielding $\angle PDK = 60°$.

All the preliminary work is done and now, to simplify matters, we will take a closer look at the diagram in Figure 2, which represents a part of Figure 1.

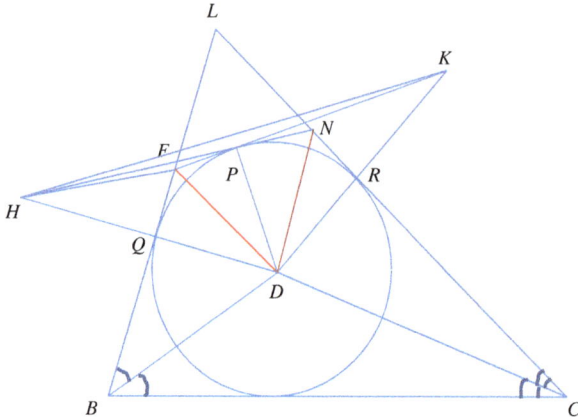

Figure 2

The sum of the opposite angles Q and R in the quadrilateral $QLRD$ is $180°$ as each of them is a right angle. Then the sum of the other two angles must be $180°$ as well:

$$\angle QDR + \angle QLR = 180°,$$

and therefore,

$$\angle QDR = 180° - \angle QLR. \tag{1}$$

In triangle BLC, $\angle L = 180° - (\angle B + \angle C) = 180° - (2\beta + 2\gamma)$.

From (*) and (1), we get that

$$\angle QDR = 180° - (180° - (2\beta + 2\gamma)) = 2\beta + 2\gamma = 120° - 2\alpha,$$

and

$$\angle QLR = 180° - (120° - 2\alpha) = 60° + 2\alpha.$$

The right triangles FQD and FPD are congruent (FD is a common hypotenuse and $DP = DQ = r$). Then

$$\angle FDQ = \frac{1}{2}\angle QDP = \frac{1}{2}(\angle QDR - \angle PDK) = \frac{1}{2}(\angle QDR - 60°)$$

$$= \frac{1}{2}(120° - 2\alpha - 60°) = 30° - \alpha. \tag{2}$$

In the isosceles triangle HDK ($HD = DK$),

$$\angle DHK = \angle DKH = \frac{1}{2}(180° - \angle HDK) = \frac{1}{2}(180° - 120° + 2\alpha) = 30° + \alpha.$$

Using (2) we now can express angle FHK in terms of α:

$$\angle FHK = \angle DHK - \angle DHF = 30° + \alpha - (30° - \alpha) = 2\alpha. \quad (3)$$

On the other hand,

$$\angle FKH = \angle DKH - \angle DKF = 30° + \alpha - 30° = \alpha. \quad (4)$$

In the triangle HFK, angle HFK is the difference of $180°$ and the sum of the other two angles: $\angle HFK = 180° - (\angle FHK + \angle FKH)$. Recalling that $\angle A = 3\alpha$ (see Figure 1) and substituting (3) and (4) into the last equality we get that

$$\angle HFK = 180° - 2\alpha - \alpha = 180° - 3\alpha = 180° - \angle A.$$

If we now go back to Figure 1, we see that in the quadrilateral $FHAK$, the angles F and A are supplementary, making it a cyclic quadrilateral. Hence, there exists a circle circumscribed about $FHAK$.

Then from (4), we get that $\angle HAF = \angle FKH = \alpha$ as angles inscribed in the circumcircle of the quadrilateral $FHAK$ and subtended by the same chord HF. Since $\alpha = \frac{1}{3}\angle A$, we conclude that AF is the trisector of angle A, and point F has to be one of the vertices of Morley's triangle.

Let's now go back once more to Figure 2 and draw HP tangent to the incircle of the triangle BLC to its intersection with CL at point N. The proof that N is the third vertex of Morley's triangle can be done in the same manner as that shown above. Four right triangles DQF, DFP, DPN, and DRN are congruent, and each angle by the vertex D is $30°$. Then in the triangle FDN, sides FD and DN are equal and $\angle FDN = 60°$, which proves the validity of the theorem, triangle FDN is equilateral.

Even though the proof is not trivial, it was based solely on basic facts learned in high school geometry. In my opinion, the importance of Morley's Theorem is in the ingenuity behind it. One might argue that its application in other areas of geometry is limited. From the practical standpoint that is perhaps true. But let's take a little different approach to that statement. How often do you hear the common perception that mathematics is boring, too abstract, difficult to comprehend, or even just useless in real life for an ordinary person? What argument against this could be more convincing than the vivid, amazingly simple geometrical phenomena presented in this chapter? It's hard to think about any other more capable of conveying the excitement and beauty of mathematics. Why should it not be used as a promotional tool to invite kids into the world of geometrical miracles, one of which was discovered relatively very recently and requires no more than the knowledge of high school mathematics?

Many years ago, I came across Hermann Schwarz's proof of Fagnano's Theorem in a mathematical journal. That problem captured my attention by the simplicity of its statement and its non-trivial solution. I spent a few days trying to get to my own proof using the least number of reflections possible. I still remember a great feeling of satisfaction when I managed to find such a solution. You can't compare that feeling to anything else. Sometimes it takes just one problem or one issue to ignite your flame of passion for the subject. In my case, Fagnano's Theorem was a spark and motivational tool in starting the geometrical journey in my life.

How exciting would it be to discover your own solution to Morley's Theorem?

It is a real challenge to find different proofs to this amazing theorem and explore generalizations derived from it. One great source for the information may be found at the website created by A. Bogomolny: Morley's Miracle, http://www.cut-the-knot.org/ triangle/morley/index.shtml. You will see an excellent selection of proofs, history, and observations regarding this unique problem.

I hope that the reader's exposure to this chapter and acquaintance with this marvelous classic problem will inspire further research and maybe even new discoveries.

With this in mind we have a fine way to close our *Geometrical Kaleidoscope*.

> *It is not knowledge, but the act of learning. Not possesion but the act of getting there, which grants the greatest enjoyment.*

> Carl Friedrich Gauss

Solutions to the Problems and Exercises

Chapter 1

Problem 4.

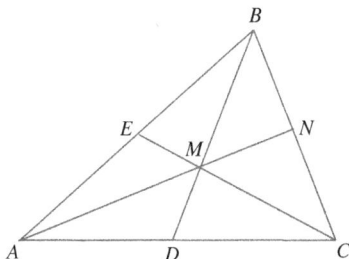

Triangle BCM consists of two triangles BMN and CMN with the area of each equal to $\frac{1}{6}$ of the area of triangle ABC. The same statement is true for the quadrilateral $AEMD$ (it consists of triangles AEM and ADM). Therefore, the area of BMC equals to the area of $AEMD$ and is $\frac{1}{3}$ of the area of the triangle ABC.

Problem 5. You may refer to the solution of Problem 9 from Chapter 6. In that problem, it is proved that the area of a triangle is $\frac{4}{3}$ of the area of the triangle formed by its medians, which would yield $S = \sqrt{q(q - m_a)(q - m_b)(q - m_c)}$ by Heron's formula.

Therefore, the area of the original triangle would be $\frac{4}{3}\sqrt{q(q - m_a)(q - m_b)(q - m_c)}$.

Problem 6.

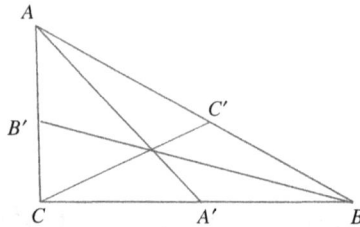

Let the lengths of the medians in triangle ABC be $AA' = m_1$, $BB' = m_2$, and $CC' = m_3$. We need to prove that AC is perpendicular to BC if

$$m_1^2 + m_2^2 = 5m_3^2. \tag{1}$$

The problem can easily be solved applying the formulas expressing the lengths of the sides of a triangle through the medians introduced in Chapter 1.

Let's denote the sides of the triangle ABC as $AB = c$, $BC = a$, $AC = b$.

Squaring both sides in the formulas derived in Chapter 1, we have

$$a^2 = \frac{4}{9}(2m_2^2 + 2m_3^2 - m_1^2),$$

$$b^2 = \frac{4}{9}(2m_1^2 + 2m_3^2 - m_2^2),$$

$$c^2 = \frac{4}{9}(2m_1^2 + 2m_2^2 - m_3^2).$$

Adding the first two expressions and substituting $m_1^2 + m_2^2 = 5m_3^2$, we have

$$a^2 + b^2 = \frac{4}{9}(2m_2^2 + 2m_3^2 - m_1^2 + 2m_1^2 + 2m_3^2 - m_2^2)$$

$$= \frac{4}{9}(m_1^2 + m_2^2 + 4m_3^2) = \frac{4}{9}(5m_3^2 + 4m_3^2) = \frac{4}{9} \cdot 9m_3^2 = 4m_3^2. \tag{*}$$

Modifying the respective expression for c^2 from the third equality, we arrive at

$$c^2 = \frac{4}{9}(2m_1^2 + 2m_2^2 - m_3^2) = \frac{4}{9}(2 \cdot 5m_3^2 - m_3^2) = \frac{4}{9} \cdot 9m_3^2 = 4m_3^2 \quad (**)$$

Comparing (*) and (**), we see that $a^2 + b^2 = c^2$. Applying the converse to the Pythagorean Theorem, we conclude that triangle ABC is the right triangle with $\angle ACB = 90°$, which is the desired result.

There is an alternative nice solution to the problem by means of vector algebra. We assume the readers are familiar with the properties of Euclidean vectors. If we manage to prove that the vector scalar product of $\overrightarrow{CA}$ by $\overrightarrow{CB}$ equals to 0, then CA is perpendicular to CB.

Notice that $\overrightarrow{AA'} = \overrightarrow{AC} + \overrightarrow{CA'} = \overrightarrow{AC} + \frac{1}{2}\overrightarrow{CB}$ and $\overrightarrow{BB'} = \overrightarrow{BC} + \overrightarrow{CB'} = \overrightarrow{BC} + \frac{1}{2}\overrightarrow{CA}$.

Square both sides of these equalities and substitute the values for the medians:

$$m_1^2 = (\overrightarrow{AC})^2 + \frac{1}{4}(\overrightarrow{CB})^2 + \overrightarrow{AC} \cdot \overrightarrow{CB},$$

$$m_2^2 = (\overrightarrow{BC})^2 + \frac{1}{4}(\overrightarrow{CA})^2 + \overrightarrow{BC} \cdot \overrightarrow{CA}.$$

After adding these equalities and combining like terms we get $m_1^2 + m_2^2 = \frac{5}{4}(\overrightarrow{AC})^2 + \frac{5}{4}(\overrightarrow{BC})^2 + 2\overrightarrow{AC} \cdot \overrightarrow{BC}$. Let's now substitute the value from (1) and multiply both sides by $\frac{4}{5}$:

$$4m_3^2 = (\overrightarrow{AC})^2 + (\overrightarrow{BC})^2 + \frac{8}{5}\overrightarrow{AC} \cdot \overrightarrow{BC}. \tag{2}$$

If we express vector $\overrightarrow{CC'}$ in terms of vectors $\overrightarrow{CA}$ and $\overrightarrow{CB}$, then $\overrightarrow{CC'} = \frac{1}{2}(\overrightarrow{CA} + \overrightarrow{CB})$. After squaring both sides (recall that $|\overrightarrow{CC'}| = m_3$) we get that

$$4m_3^2 = (\overrightarrow{CA})^2 + (\overrightarrow{CB})^2 + 2\overrightarrow{CA} \cdot \overrightarrow{CB}.$$

Comparing the last equality and (2) and canceling like terms yields

$(\overrightarrow{AC})^2 + (\overrightarrow{BC})^2 + \frac{8}{5}\overrightarrow{AC} \cdot \overrightarrow{BC} = (\overrightarrow{CA})^2 + (\overrightarrow{CB})^2 + 2\overrightarrow{CA} \cdot \overrightarrow{CB}$, from which $\overrightarrow{AC} \cdot \overrightarrow{BC} = 0$. Then AC must be perpendicular to BC, and we are done.

Problem 7.

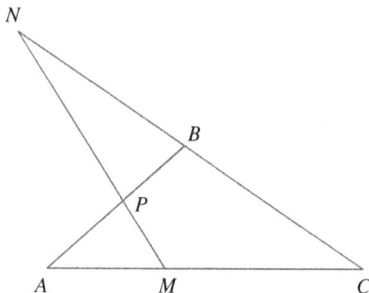

Given $AM = \frac{1}{2}MC$ and $BC = BN$, we need to find the ratio AP:PB.

The application of the properties of a center of gravity of a system of points allows getting a relatively simple and elegant solution. Assume that equal weights of 1 pound are placed at point C and at point N and 2 pounds at point A. We get the system of points $1C$, $1N$, and $2A$. Denote the center of gravity of that system by O. Let's figure out where O should lie. First, the center of gravity of points A and C must be M. Indeed, we have the system of two points $2A$ and $1C$ and $AM = \frac{1}{2}MC$. Therefore, by Archimedes' Law of the Lever, M is the center of gravity of that system of two points. If we concentrate all masses at M, we will get the new system consisting of point $3M$ and point $1N$. The center of gravity of that system, O, has to lie on NM, and by Archimedes' Law of the Lever, $NO = 3OM$. On the other hand, if we concentrate the masses from points C and N at point B (which is the midpoint of segment NC and therefore, is the center of gravity of $1N$ and $1C$), then the system of the two points $2A$ and $2B$ must have the same point O as its center of gravity. O has to lie on AB. We proved that the center of gravity of our system of points must be the intersection of MN and AB, which means O coincides with P. After the application of Archimedes' Law of the

Lever to points $2A$ and $2B$, we get that

$$AP{:}PB = 2{:}2 = 1.$$

Problem 8. The greatest is the side with the length 13. By applying the formula from the corollary to Appollonius's Theorem, we get

$$m_a^2 = \frac{1}{4}(2b^2 + 2c^2 - a^2) = \frac{1}{4} \cdot (2 \cdot 11^2 + 2 \cdot 12^2 - 13^2),$$

$$m_a^2 = \frac{1}{4} \cdot 361,$$

$$m_a = 9.5.$$

Chapter 2

Problem 9. Draw the perpendiculars from accessible vertices to opposite sides of the triangle. The point of their intersection is the orthocenter. Draw a perpendicular through the orthocenter to the third side. It must contain the third altitude and, therefore, will pass through the third inaccessible vertex.

Problem 10.

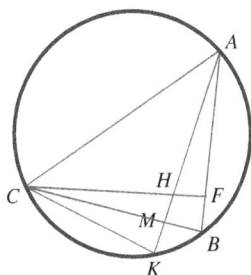

Assume A is the given point on the circle and H is the given point inside it. Draw line AH to its intersection with the circle at K. Find M, the midpoint of HK. Draw the line at M perpendicular to AK to its intersection with the circle at points C and B. ABC is the triangle with all its vertices lying on a given circumference. By Theorem 2 from Chapter 3, H is the orthocenter of the triangle ABC.

Problem 11.

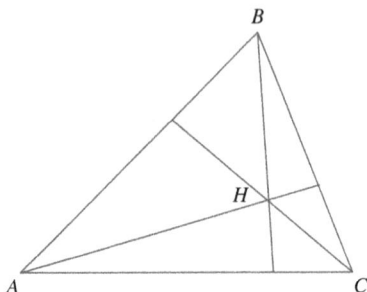

Assume R is the radius of the circumcircle of ABC and r_1 is the radius of the circumcircle of AHC. For the proof, we need to use the formula for the radius of a circumcircle of a triangle through a side and the sine of an opposite angle (by the Law of sines):

In triangle ABC, $\dfrac{AC}{\sin \angle ABC} = 2R$.

In triangle AHC, $\dfrac{AC}{\sin \angle AHC} = 2r_1$.

In Theorem 1, we proved that $\angle ABC + \angle AHC = 180°$.

Since $\sin \angle ABC = \sin(180° - \angle AHC) = \sin \angle AHC$ then $R = r_1$. In the same way, you can get that $r_2 = r_3 = r_1 = R$.

Chapter 3

Problem 2.

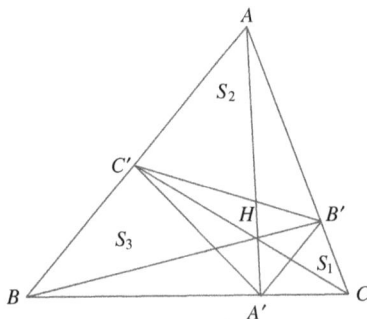

By Theorem 1, triangles $A'B'C$, $B'AC'$, and $C'BA'$ are all similar to triangle ABC with the ratios $\cos C$, $\cos A$, and $\cos B$, respectively.

If the areas of the triangles are S_1, S_2, S_3, and S, respectively, then by the Theorem of the ratios of the areas of similar triangles we get

$$S_1/S = \cos^2 C$$
$$S_2/S = \cos^2 A$$
$$S_3/S = \cos^2 B.$$

Adding yields $S_1 + S_2 + S_3 = S(\cos^2 C + \cos^2 A + \cos^2 B)$.

With a few trigonometric manipulations, it's not hard to prove that

$$\cos^2 C + \cos^2 A + \cos^2 B = 1 - 2\cos C \cos A \cos B.$$

To prove it the following well-known trigonometric formulas will be used:

$2\cos^2 \alpha = 1 + \cos 2\alpha$ and $\cos \alpha + \cos \beta = 2\cos\left(\frac{\alpha+\beta}{2}\right)\cos\left(\frac{\alpha-\beta}{2}\right)$ (for any angles α and β). Remember also that $\angle A + \angle B + \angle C = 180°$, from which $\angle C = 180° - (\angle A + \angle B)$.

We have

$\cos^2 A + \cos^2 B + \cos^2 C$

$= (1 + \cos(2A))/2 + (1 + \cos(2B))/2 + (1 + \cos(2C))/2$

$= (3 + \cos(2A) + \cos(2B) + \cos(2C))/2$

$= 1 + (1 + \cos(2A) + \cos(2B) + \cos(2C))/2$

$= 1 + (\cos(2A) + \cos(2B) + 1 + \cos(360° - 2(A + B)))/2$

$= 1 + (2\cos(A + B)\cos(A - B) + 1 + \cos 2(A + B))/2$

$= 1 + (2\cos(A + B)\cos(A - B) + 2\cos^2(A + B))/2$

$= 1 + \cos(A + B)\cos(A - B) + \cos^2(A + B)$

$= 1 + \cos(A + B)(\cos(A - B) + \cos(A + B))$

$= 1 + \cos(180° - C) \cdot 2\cos A \cos B = 1 - 2\cos C \cos A \cos B.$

Notice that the area of the triangle $A'B'C'$ equals the difference of S and $(S_1 + S_2 + S_3)$, we have $S_{A'B'C'} = S - (S_1 + S_2 + S_3) = S - S(\cos^2 C + \cos^2 A + \cos^2 B) = S - S(1 - 2\cos C \cos A \cos B) =$

$2S \cos C \cos A \cos B$. This yields the desired result
$S_{A'B'C'}/S = 2 \cos C \cos A \cos B$.

Problem 3. It has to be proved that $R = 2r$, where R is the radius of the circumcircle of ABC and r is the radius of the circumcircle of its orthic triangle $A'B'C'$.

We refer to the diagram used for Problem 2. Let H be the orthocenter of the triangle ABC.

In the corollary to Theorem 2, it was proved that

$$\angle A'C'B' = 180° - 2\angle ACB. \tag{1}$$

From the similarity of the triangles ABC and $A'B'C$ follows

$$A'B' = AB \cos C. \tag{2}$$

In the triangle ABC, the radius of its circumcircle

$$R = \frac{AB}{2 \sin C}. \tag{3}$$

Expressing similarly r in terms of $A'B'$ and angle $A'C'B'$ in the triangle $A'B'C'$ and substituting values from (1), (2), and (3), and using $\sin 2\alpha = 2 \sin \alpha \cos \alpha$, we get

$$r = \frac{A'B'}{2 \sin \angle A'C'B'} = \frac{AB \cos C}{2 \sin(180° - 2C)} = \frac{AB \cos C}{2 \sin 2C}$$
$$= \frac{AB \cos C}{4 \sin C \cos C} = \frac{AB}{4 \sin C} = \frac{1}{2}R.$$

Problem 4. The orthic triangle of a given triangle can be an example of the triangle with the required sides.

Problem 7.

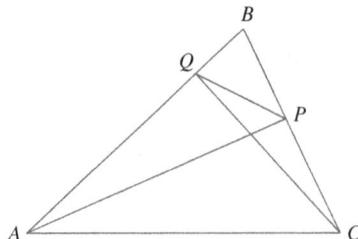

Given $AP \perp BC$, $CQ \perp AB$, $P_{ABC} = 15$, $P_{BPQ} = 9$, and the radius of the circumcircle of triangle BPQ, $r = \frac{9}{5}$, we need to find AC.

Triangles ABC and PBQ are similar with the ratio $\cos B$.

Then $\frac{P_{BPQ}}{P_{ABC}} = \cos B = \frac{9}{15} = \frac{3}{5}$.

Recalling that $\sin^2 B + \cos^2 B = 1$, we get that

$$\sin B = \sqrt{1 - \left(\frac{3}{5}\right)^2} = \frac{4}{5}.$$

In the triangle PBQ, $\frac{PQ}{\sin B} = 2r$, from which

$PQ = 2r \cdot \sin B = \frac{18}{5} \cdot \frac{4}{5} = \frac{72}{25}$.

Once again, from the similarity of the triangles ABC and PBQ it follows that

$$AC = \frac{PQ}{\cos B} = \frac{72}{25} \cdot \frac{5}{3} = \frac{24}{5}.$$

Problem 8. The result is achieved by applying the Theorem of ratios of the areas of similar triangles PBQ and ABC and using the Law of sines, from which $R = \frac{AC}{2\sin B}$.

The answer is $R = 4.5$.

Chapter 4

Problem 7. You might look at the solution of Problem 3 in Chapter 8.

Problem 8. First, find the hypotenuse in the triangle ABC by the Pythagorean Theorem. Then the result follows directly from the formula derived in Theorem 2.

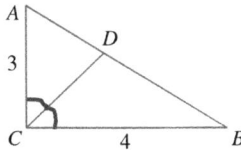

$AB^2 = AC^2 + CB^2 = 9 + 16 = 25$. Then $AB = 5$.

$$CD = \frac{2 \cdot 3 \cdot 4}{3 + 4} \cdot \cos 45° = \frac{12\sqrt{2}}{7}.$$

Problem 9. Given $AB = BC = a$ and $AC = b$, and CM and AN are the angle bisectors in the triangle ABC, find MN.

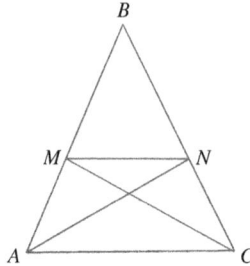

Triangle ABC is isosceles. Triangles AMC and CAN are congruent by ASA property (AC is a common side, $\angle MAC = \angle NCA$ and $\angle MCA = \angle NAC$). Then triangles AMN and CNM are congruent as well (by SSS property). Therefore, $\angle CMN = \angle ANM$, which immediately implies that $\angle CMN = \angle MCA = \angle NAC = \angle ANM$, and hence $MN \parallel AC$. From the last fact, we conclude that triangles MBN and ABC are similar, so

$$MN/AC = BM/AB. \qquad (1)$$

CM is the bisector of angle ACB, so by Theorem 1,

$$MA/MB = AC/CB. \qquad (2)$$

$MA = AB - MB$; substituting it in (2), we get
$(AB - MB)/MB = AC/CB$, from which

$$AB/BM = 1 + AC/CB,$$
$$AB/BM = 1 + (b/a) = (a + b)/a.$$

Then $BM/AB = a/(a + b)$. Substituting this expression in (1) gives $MN/b = a/(a + b)$, from which $MN = ab/(a + b)$.

Chapter 6

Problem 6.

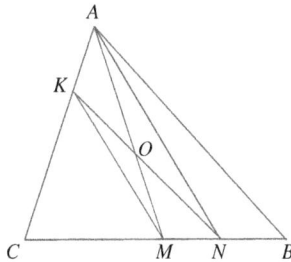

Draw the median AM. Connect points K and M. Then draw $AN \parallel KM$, where N lies on BC. The final step is to draw KN.

By construction, $AKMN$ is a trapezoid with diagonals KN and AM. As was proved in Problem 4 from Chapter 6, the area of the triangle AOK equals the area of the triangle MON. AM is a median in the triangle ABC, so $S_{CAM} = \frac{1}{2}S_{ABC}$. Because $S_{CKN} = S_{CKOM} + S_{OMN} = S_{CKOM} + S_{AOK} = S_{CAM} = \frac{1}{2}S_{ABC}$, we see that KN is the desired line. KN divides ABC into two parts of equal area.

Problem 7.

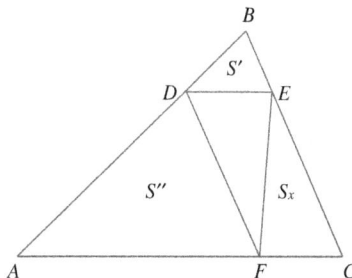

Given $DF \parallel BC$ and $DE \parallel AC$, and $S_{ADF} = S''$ and $S_{BDE} = S'$, find S_x, the area of CEF.

Let's denote by S the area of the triangle ABC. $DECF$ is a parallelogram by construction. Hence, the areas of triangles DEF and

EFC are equal.

$$\text{Therefore, } S = S' + S'' + 2S_x. \tag{1}$$

Triangles ADF, DBE, and ABC are all similar to each other.
Thus $\frac{\sqrt{S'}}{\sqrt{S}} = \frac{DE}{AC}$ and $\frac{\sqrt{S''}}{\sqrt{S}} = \frac{AF}{AC}$.

Adding and noticing that $FC = DE$ (opposite sides of a parallelogram), we get $\frac{\sqrt{S'}+\sqrt{S''}}{\sqrt{S}} = \frac{DE+AF}{AC} = \frac{FC+AF}{AC} = 1$, leading to $\sqrt{S'} + \sqrt{S''} = \sqrt{S}$. Squaring yields

$$S = S' + S'' + 2\sqrt{S'S''}.$$

Substituting in (1) leads to:

$$S_x = \sqrt{S'S''}.$$

Problem 8. The triangle formed by the midlines will be similar to the original triangle with the ratio $\frac{1}{2}$.

$$\frac{S_{A'B'C'}}{S_{ABC}} = \left(\frac{1}{2}\right)^2 = \frac{1}{4}.$$

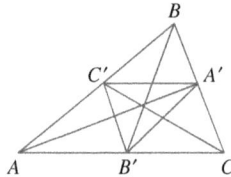

Problem 9.

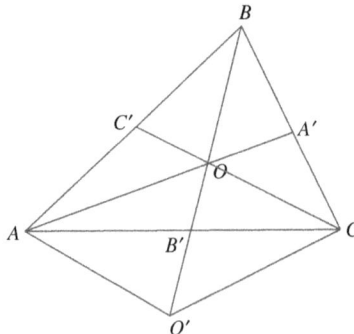

Let AA', BB', and CC' be the medians of the triangle ABC, and O is its centroid, the point of intersection of the medians.

Rotate point O by $180°$ about B' and get its image point O' on the extension of OB'. Then by construction, the diagonals of the quadrilateral $AOCO'$ are divided in half by the point of their intersection, which means it is a parallelogram. Hence $CO' = AO$ as opposite sides of a parallelogram.

By a property of medians, $CO = \frac{2}{3}CC'$ and $AO = \frac{2}{3}AA'$. Also, $OB' = \frac{1}{2}OB$ and by construction, $O'B' = OB'$. Then $OO' = OB$. We conclude that all three sides of the triangle $O'OC$ are $\frac{2}{3}$ of their respective medians of triangle ABC.

The area of the triangle $OB'C$ is $\frac{1}{6}$ of the area of the triangle ABC, which was proved in Chapter 1. By noting that the area of the triangle $OO'C$ is twice the area of the triangle $OB'C$ (the median CB' divides OCO' into two triangles of equal areas), we get $S_{OO'C} = 2 \cdot \frac{1}{6}S_{ABC} = \frac{1}{3}S_{ABC}$. Our triangle $O'OC$ is similar to the triangle formed by the medians of the triangle ABC with the ratio $\frac{2}{3}$. By the Theorem of ratios of the areas of similar polygons, the ratio of their areas is $\left(\frac{2}{3}\right)^2 = \frac{4}{9}$, therefore

$$\frac{S_{OO'C}}{S} = \frac{4}{9}, \text{ or } S = \frac{9}{4}S_{OO'C} = \frac{9}{4} \cdot \frac{1}{3}S_{ABC} = \frac{3}{4}S_{ABC}.$$

Problem 10.

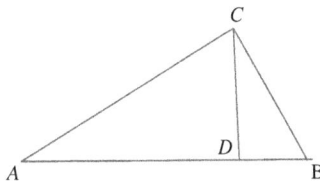

Denote by S the area of triangle ABC, S_1 the area of triangle CDB, and S_2 the area of triangle ADC.

Triangles ACB, ADC, and CDB are all similar, having the same three angles. Therefore,

$$\frac{S_1}{S} = \frac{l_1^2}{l^2},$$

$$\frac{S_2}{S} = \frac{l_2^2}{l^2}.$$

Adding yields

$$\frac{S_1}{S} + \frac{S_2}{S} = \frac{l_1^2}{l^2} + \frac{l_2^2}{l^2}, \text{ so } \frac{S_1 + S_2}{S} = \frac{l_1^2 + l_2^2}{l^2}.$$

Because $S_1 + S_2 = S$, we get $l_1^2 + l_2^2 = l^2$.

Chapter 7

Problem 7.

The problem will be solved if we manage to draw the straight line at O that intersects the sides of the given angle at points K and D such that O is the midpoint of KD. The triangle AKD will have the smallest area of all the triangles cut from the angle by straight lines passing through O.

Let's do the construction first. After that we will prove the statement above.

You can use the construction suggested in Solution 3 from the chapter "Session of One Interesting Problem". Here we show another way of doing it.

Find the image of A in a central symmetry with the center O, point B on line AO, such that $AO = OB$. Draw two lines at B parallel to the sides of the given angle and denote K and D the points of their intersection with the angle's sides.

$AKBD$ is a parallelogram by construction. O is the point of intersection of its diagonals AB and KD, therefore, $KO = OD$.

Now we pick a random point M on a side AK and draw MO to its intersection with AD at N. Let's prove that area of the triangle AMN is greater than area of the triangle AKD.

Denote by L the point of intersection of MN and DB. Triangles MOK and LOD are congruent (ASA property). Therefore, their areas are equal:

$$S_{MOK} = S_{LOD}.$$

By comparing the areas of triangles AKD and AMN, we see that

$$S_{AKD} = S_{AMOD} + S_{MOK} \text{ and}$$
$$S_{AMN} = S_{AMOD} + S_{NOD}.$$

The quadrilateral $AMOD$ is a common part of both triangles, so basically, we have to compare the areas of triangles MOK and NOD. The area of NOD is greater than the area of MOK by the area of the triangle DLN. Then $S_{AKD} < S_{AMN}$ and the proof is complete. Triangle AKD has the smallest area of all such triangles.

Problem 8.

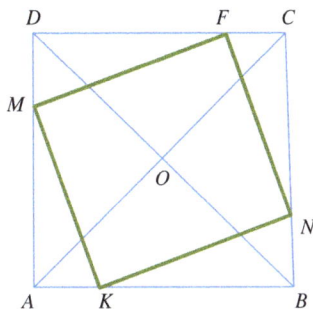

Rotate M about O (the point of intersection of the diagonals of $ABCD$) through an angle of $90°$. In that rotation, AD rotates into DC, DC into CB, and CB into BA. M will be rotated into F on side DC such that $DF = AM$. The image of F will be N on BC such that $CN = DF$ and the image of N will be K on AB such that $KB = CN$.

To summarize, MF is rotated into FN, FN into NK, and NK into KM.

Since rotation preserves the distances between points, $MF = FN = NK = KM$. The rotation was through a $90°$ angle, therefore the angles between respective images are all $90°$, i.e., in the quadrilateral $MFNK$, angles M, F, N, and K are right angles. Thus $MFNK$ has all equal sides and all equal angles of $90°$. Therefore, it is a square.

Problem 9.

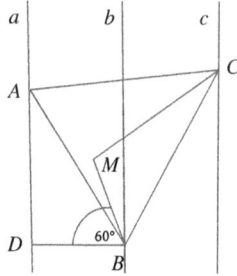

It is given that $a \parallel b \parallel c$. We have to construct an equilateral triangle with vertices on lines a, b, and c.

Select any point B on line b. Draw BD perpendicular to a (D is on the line a).

Rotate line a by $60°$ clockwise around B. Denote by M the image of D. Then $BD = BM$ and $\angle DBM = 60°$. Draw the line at M perpendicular to BM intersecting c at C. The last step is to draw the circle with center B and radius BC. It intersects a at A. The triangle ABC is an equilateral triangle and its vertices lie on the given parallel lines.

Let's prove it.

Since ABC is an isosceles triangle by construction, then to prove that ABC is an equilateral triangle it suffices to prove that $\angle ABC = 60°$.

The right triangles ADB and CMB are congruent ($DB = BM$ and $AB = BC$ by construction). Because we rotated line a by $60°$ about B and M was the image of D, we get that C is the image of A. It implies that the triangle ADB was rotated into the triangle CMB. The angle between the respective segments is preserved and is equal to $60°$. Hence the angle ABC is $60°$, which completes the proof. ABC is the desired equilateral triangle with its vertices lying on the three given parallel lines.

Appendix

Basic Selected Definitions, Formulas, and Theorems

- By definition, two triangles are congruent if their corresponding sides are equal in length, and their corresponding angles are equal in measure.
- Two triangles are congruent if the sides of the first triangle are congruent to the corresponding sides of the second triangle — *SSS property*.
- Two triangles are congruent if two pairs of corresponding sides of two triangles are congruent, and the angles between them are congruent as well — *SAS property*.
- Two triangles are congruent if two angles and the side between them in one triangle are congruent to the corresponding two angles and the side between them in other triangle — *ASA property*.
- The sum of the interior angles of any triangle is $180°$.
- If a pair of alternate interior angles formed by a transversal of two straight lines are congruent, the lines are parallel. Converse holds as well.

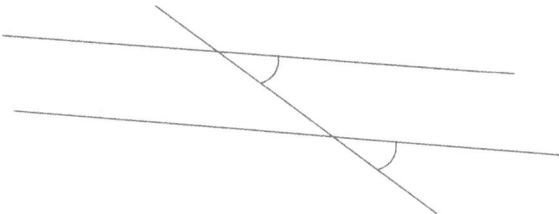

- The sum of the lengths of any two sides of a triangle is greater than the length of the third side — *The triangle inequality theorem.*
- The segment joining the midpoints of two sides of a triangle (midline) is parallel to the third side and half as long as the third side — *The midpoint theorem.*
- If two or more parallel lines are intersected by two self-intersecting lines, then the ratio of the line segments of the first intersecting line is equal to the ratio of the similar line segments of the second intersecting line — *Thales's theorem.*

$$\frac{m}{n} = \frac{x}{y}$$

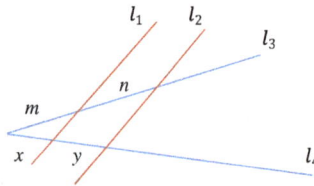

- A greater angle of a triangle is opposite a greater side.
- A greater side of a triangle is opposite a greater angle.
- Two triangles are similar if two angles of one triangle are congruent to two corresponding angles of the other triangle.
- Two triangles are similar if an angle of one triangle is congruent to an angle in the other triangle, and the lengths of the sides forming these angles are proportional.
- In a right triangle, the square of the hypotenuse (side opposite to the right angle) is equal to the sum of the squares of the legs: $c^2 = a^2 + b^2$ – *the Pythagorean Theorem.* Converse holds as well.
- The length of the altitude on the hypotenuse of a right triangle is the geometrical mean between the lengths of the segments of the hypotenuse.

$$CD = \sqrt{AD \cdot DB} \quad \text{or} \quad h = \sqrt{h_a \cdot h_b}$$

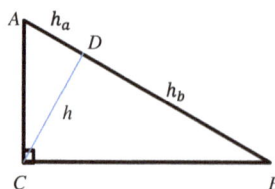

- In a triangle with sides of the lengths a, b, and c, and angle γ between sides a and b, the following equality holds:
  $a^2 + b^2 - 2ab \cdot \cos\gamma = c^2$ – *the Law of cosines.*

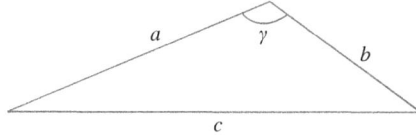

- In a triangle with sides of the lengths a, b, and c, and angles α, β, and γ respectively opposite to these sides, and R is the radius of the triangle's circumcircle, the following equality holds:

$$\frac{a}{\sin\alpha} = \frac{b}{\sin\beta} = \frac{c}{\sin\gamma} = 2R - \textit{the Law of sines.}$$

- Formulas to calculate the area of a triangle:
  $S = \frac{1}{2}ah_a$, where a is the base and h_a is the altitude drawn to this base.
  $S = \frac{1}{2}ab \cdot \sin\gamma$, where γ is the angle between the sides a and b.
  $S = \sqrt{p(p-a)(p-b)(p-c)}$, where a, b, and c are the sides and p is the semi-perimeter of a triangle, $p = \frac{1}{2}(a+b+c)$ — *Heron's formula.*
- A parallelogram is a quadrilateral with both pairs of opposite sides parallel.
- A trapezoid is a quadrilateral with only one pair of opposite sides parallel.
- The opposite sides and angles in a parallelogram are congruent. Converse holds as well.
- Pairs of consecutive angles in a parallelogram are supplementary (their sum is $180°$).
- The diagonals of a parallelogram bisect each other. Converse holds as well.
- Formulas to calculate the area of a parallelogram:
  $S = ah_a$, where a is the base and h_a is the altitude drawn to this base.
  $S = ab \cdot \sin\gamma$, where γ is the angle between the sides a and b.
- Formula to calculate the area of a trapezoid:
  $S = \frac{1}{2}(a+b)h$, where a and b are the bases and h is the altitude.
- A line perpendicular to a radius at a point on the circle is tangent to the circle at that point.

- A line perpendicular to a tangent line at the point of tangency with a circle passes through the center of the circle.
- Two tangent segments to a circle from an external point are congruent.

$$AM = AN$$

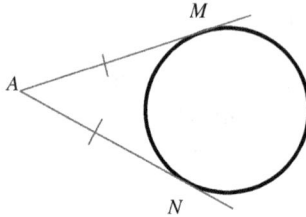

- The center of the circumscribed circle of a triangle is the point of intersection of the perpendicular bisectors of the sides of the triangle.
- The center of the inscribed circle of a triangle is the point of intersection of the interior angle bisectors of the triangle.
- An angle inscribed in a circle is half of the central angle that subtends the same arc on the circle — *the Inscribed angle theorem.*

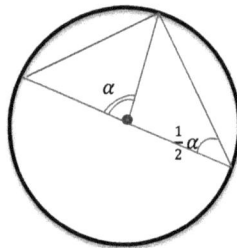

- An inscribed angle subtended by the diameter, is a right angle — direct consequence from *the Inscribed angle theorem.*
- All angles inscribed in a circle and subtended by the same chord are congruent — direct consequence from *the Inscribed angle theorem.*
- Given a secant l intersecting the circle at points M and N and tangent t touching the circle at point T and given that l and t intersect at point P, the following equality holds:
  $PT^2 = PM \cdot PN$ — the *Tangent-secant theorem.*

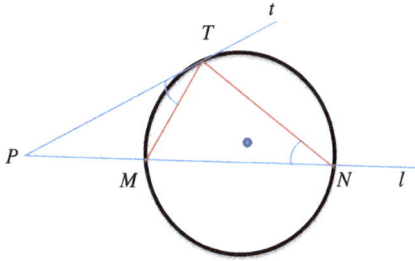

- The measure of the angle formed by two chords that intersect inside the circle is equal to half the sum of the chords' intercepted arcs —*Angles of intersecting chords theorem*.
- For two chords that intersect in a circle, the products of the lengths of the line segments on each chord are equal — *Intersecting chords theorem*.
- If a convex quadrilateral is inscribable in a circle then the product of the lengths of its diagonals is equal to the sum of the products of the lengths of the pairs of opposite sides — *Ptolemy's theorem*.

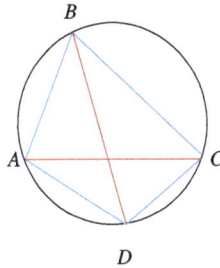

$$AC \cdot DB = AB \cdot DC + DA \cdot BC$$

- A Euclidean vector is a geometric object defined by its length and a direction and denoted $\overrightarrow{AB} = \vec{a}$

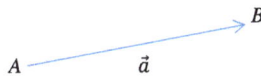

- Additions of vectors :

$$\overrightarrow{AB} + \overrightarrow{BC} = \overrightarrow{AC}$$

Triangle rule:

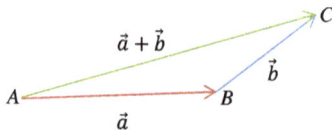

Parallelogram rule:

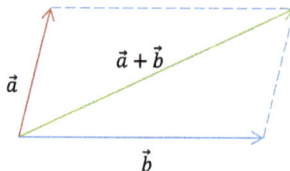

- The *dot product* (or a *scalar product*) of two non-zero vectors $\vec{a}$ and $\vec{b}$ is the product of the length of $\vec{a}$ by the length of $\vec{b}$ and by the cosine of the angle between them: $\vec{a} \cdot \vec{b} = \|\vec{a}\| \cdot \|\vec{b}\| \cdot \cos\gamma$, where $\|\vec{a}\|$ is the length of the vector $\vec{a}$, $\|\vec{b}\|$ is the length of the vector $\vec{b}$, and γ is the angle these vectors form.

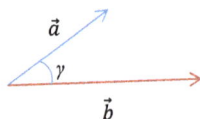

References

Bogomolny, A., *Morley's Miracle*, http://www.cut-the-knot.org/triangle/ Morley/index.

Boyer, C.B., *A History of Mathematics* (2nd edition), John Wiley & Sons, Inc., 1991.

Coxeter, H.S.M, *Introduction to Geometry*, John Wiley & Sons, New York, 1961.

Coxeter, H.S.M. and S.L. Greitzer, *Geometry Revised*, MAA, 1967.

Ed. Sandifer, The Euler Line, *How Euler Did It*, MAA Online, January 2009.

Grunbaum, B., Is Napoleon's Theorem *Really* Napoleon's Theorem?, *American Mathematical Monthly* 119(6) (2012) 495–501.

Johnson, A., *Famous Problems and Their Mathematician*, Teacher's Idea Press, Greenwood Village, CO, 1999.

Johnson, R.A., *Advanced Euclidean Geometry*, Dover Publications, 2007.

Kohn, E., *Geometry*, Cliffs Notes, Inc. 1994.

Kordemsky, B., What's the best answer?, *Quantum*, July/August (1996).

Loomis, E.S., *The Pythagorean Proposition*, NCTM, Reston, VA, 1968.

Oakley, C.O. and Baker, J.C., The Morley Trisector Theorem, *Publ. Math. IHES*, 88 (1988) 43–46.

Ogilvy, C.S., *Excursions in Geometry*, Dover, 1990.

Polya, G., *Mathematical Discovery*, John Wiley & Sons, 1997.

Posamentier, A. and Lehmann I., *The Secretes of Triangles: A Mathematical Journey*, Prometheus Books, 2012.

Posamentier, A. and Salkind, C., *Challenging Problems in Geometry*, Dover, 1996.

Pritsker, B., A Pivotal Approach, *Quantum*, May/June (1996).

Pritsker, B., Auxiliary Elements in Problem Solving, *Mathematics and Informatics Quarterly*, 8(2) (1998).

Pritsker, B., Bisector of a Triangle and its Properties, *NY State Mathematics Teachers' Journal*, 46(3) (1996).

Pritsker, B., Constructions-Siblings, *Mathematics and Informatics Quarterly*, 9(3) (1999).

Pritsker, B., Exploring every Angle, *Quantum*, March/April (2001).

Pritsker, B., Some Properties of the Orthic Triangle, *Mathematics Competition*, 9(2) (1996).

Pritsker, B., The Area of a Quadrilateral, *NY State Mathematics Teachers' Journal*, 45(3) (1995).

Pritsker, B., The Orthocenter of a Triangle and Some of its Properties, *Journal of Recreational Mathematics*, 26(4) (1994).

Silvester, J.R., *Geometry: Ancient & Modern*, Oxford University Press, 2001.

Stonebridge, B., A simple geometric proof of Morley's trisector theorem, *Math. Spectrum*, 42 (2009) 2–4.

Бланк, М.Б. and Болтянский, В.Г., *"Применение понятия центра масс на факультативных и кружковых занятиях"*, Математика в школе, 2 (1984) c.45–50 (in Russian).

В.Гусев, В. Литвиненко, and Мордкович, А., *Практикум по решению математических задач*, Москва "Просвещение", 1985 (in Russian).

Габович, И., *Алгоритмический подход к решению геометрических задач*, Киев, Радянська школа, 1989 (in Russian).

Горделадзе, Ш., Кухарчук, М., and Яремчук, Ф., *Збірник Конкурсних Задач з Математики*, Київ, Вища Школа, 1988 (in Ukrainian).

Михайловський, В., Ядренко, М., Призва, Г., and Вишенський, В., *Збірник задач республиканських математичних олімпіад*, Київ, Вища школа, 1979 (in Ukrainian).

Прицкер, Б., *Площадь четырехугольника*, Математика в школе, 4 (1990) (in Russian).

Сканави, М.И., *Сборник конкурсных задач по математике для поступающих во втузы*, Москва "Высшая Школа", 1980 (in Russian).

Чистяков, В., *Старинные задачи по элементарной математике*, Минск Вышэйшая школа, 1978 (in Russian).

Index

A

altitude, 1, 4–5, 15–28, 32–35, 37–38, 43–44, 46–47, 55, 74–75, 77–79, 82, 90, 101, 103–104, 111, 119, 125, 128–129, 134, 140, 153, 166–167

AM–GM inequality, 67

angle bisector, 1, 45–58, 60, 90, 98–100, 103, 119, 126–127, 145, 158, 168

angle trisector, 144

Angles of Intersecting Chords Theorem, 169

Appollonius's Theorem, 3, 153

Archimedes, 9–10

  Archimedes' Law of the Lever, 9–10, 12, 152

area of a convex quadrilateral, 13, 38, 57–58, 62, 69, 101

area of a parallelogram, 62, 167

area of a trapezoid, 62, 80–81, 87, 100–101, 134, 167

area of a triangle, 4–5, 22–26, 38, 49, 57, 60–61, 66, 74–76, 78–79, 81, 87, 100–102, 116, 132, 139, 149, 159, 161, 163, 167

auxiliary element, 42–43, 93–95, 97, 100, 102, 104, 106, 126, 128, 135

B

bisector of the angle, *see also* angle bisector, 33, 44, 114

Brahmagupta, 64

  Brahmagupta's Formula, 58, 65–66, 72

Bretschneider, Carl Anton, 62

  Bretschneider's Formula, 62

Brianchon, Charles Julien, 26

C

center of gravity, *see also* centroid, 5–6, 8–12, 152

center of mass, *see also* centroid, 5–6, 9, 11–12, 118

central symmetry, *see also* half-turn, 86, 88, 162

centroid, 1–3, 5, 10–11, 13, 16–17, 28, 91, 119, 123, 127–130, 161

Ceva, Giovanni, 115, 118

  Ceva's Theorem, 109, 116–120

  cevians, 116, 118–120, 126–127

circumcenter, 16–17, 27–28, 119, 122, 128–130

circumcircle, 19–22, 25, 28–29, 34, 36, 38, 44, 50, 59, 99, 104–105, 122, 147, 154, 156–157, 167

circumradius, 25–26

  Circumradius Theorem, 25

www.ingramcontent.com/pod-product-compliance
Lightning Source LLC
Chambersburg PA
CBHW061252220326
41599CB00028B/5617